£12.99

WHO
INVENTED
WHAT
WHEN

David Ellyard

D1347816

I dedicate this book to my loving and ever-supportive family, Sue, Rachel, James, and Sam, with a special thanks to James for all his assistance with research.

Published in Australia in 2006 by
New Holland Publishers (Australia) Pty Ltd
Sydney • Auckland • London • Cape Town

14 Aquatic Drive Frenchs Forest NSW 2086 Australia
218 Lake Road Northcote Auckland New Zealand
86 Edgware Road London W2 2EA United Kingdom
80 McKenzie Street Cape Town 8001 South Africa

10 9 8 7 6 5 4 3 2

National Library of Australia Cataloguing-in-Publication Data:

    Ellyard, David, 1942- .
    Who invented what when.

    Bibliography.
    Includes index.
    ISBN 9 78174110 4882 (pbk.).

    1. Discoveries in science - Encyclopedias. 2. Science - Encyclopedias. I. Title.

503

Publisher: Martin Ford
Project Editor: Yani Silvana
Designer: Tania Gomes
Production: Monique Layt
Printer: Griffin Press, Adelaide

Contents

Introduction

*W*ho Invented What When introduces you to the technologies, inventions and innovations that have changed the lives of ordinary people in Western societies since 1500. It is a companion to *Who Discovered What When* (New Holland 2005), which covers scientific discoveries and concepts, though science and technology are closely linked, especially since the nineteenth century.

Whereas science is driven by curiosity, by the human need to *know* and understand, technology is powered by our need to survive and thrive, by what we need (or want) to do regarding food, shelter, warmth, transport, communications, staying healthy, being entertained, learning new things.

As *Who Discovered What When* did with science, this book places key inventions and innovations in a context of time, place and circumstance. You will meet the people involved, and discover their ingenuity and persistence, their triumphs and tragedies. Technology is about people as much as things, the people who invented and the people who benefited from the inventions, and so is this book. To help you explore it, we have provided indexes of 'Who' and 'What'. The 'When' is covered both by the arrangement of the stories by date, and by 'Timelines' just before the indexes, which charts the main steps in various key areas of technology over the last 500 years and guides you to the stories.

Many more inventions have made their mark on the planet than could fit in this book. In making my selection, I was guided firstly by what I thought readers would expect to find here. However, I deliberately added things that will, I hope, surprise you. I concentrated on new technologies that have made a real impact, on *innovations* rather than just *inventions*. I tried to keep a reasonable balance between technologies for different purposes. And I looked for good human stories.

You can, I think, distil some insights into innovation from the stories recounted here. As Thomas Edison famously remarked: 'Invention (meaning innovation) is 1 per cent inspiration and 99 per cent perspiration'. Having the bright idea, even identifying the need, are only the first steps. Getting the technology into peoples' lives and workplaces, say through the market, is the real struggle, and many bright ideas do not make it.

Technology takes time to have an impact, a couple of decades at least. That is why you will not find much in this book from the 1990s or later. Many of the inventions made then are yet to shape our lives. The path from invention to innovation can be littered with technical challenges. Despite promising technology and a realistic market, getting to market can take a long time. The fuel cell fits that description.

While many inventors, perhaps most, gained a good reward from their 'inspiration plus perspiration' in terms of money or recognition (Joseph Glidden, the inventor of barbed wire, was one of the richest men in America), some did not. Some were cheated, a few died in poverty and obscurity. Consider Denis Papin or Henry Cort. Think on the sad end of Nicolas Leblanc and Edwin Armstrong, both so destroyed by fate they killed themselves.

While inventions link an available process with a perceived need, that need may be unrecognised except by the inventor. No one was asking for email when that was devised. Some inventions are, at least initially, solutions in search of a problem—the laser for one. Some are pure serendipity, devised while looking for something else. William Perkin was trying to make synthetic quinine when he discovered his famous purple dye. Radar in Britain began as a 'death ray'.

Much invention is cumulative, even more than in science. Someone makes a start; someone else adds to it; other people bring in new ideas or technology. It becomes difficult to say who the inventor was. The bicycle is an example. The names associated with inventions in this book are usually those who got the ball rolling or made the most significant contribution, but that is just my judgment. From the first idea to successful innovation can take centuries. Again the associated date is the one that seemed to me most significant—marking the first stirrings, the big advance, the successful outcome.

Juxtaposing technology and need generates inventions 'whose time as come', and that can be invented independently and almost simultaneously by two or more people (for example, Alexander Bell, Elisha Grey and the telephone; or Joseph Swan, Thomas Edison and the light globe).

The earliest practitioners often worked alone, often in secret, spending their own money and trying to manage single-handedly all the steps from invention to innovation. In the last century, much more new technology has come from large laboratories run by big corporations, as Edison pioneered. The inventor can concentrate on the creative end, leaving others to handle manufacturing and marketing.

One thing is unchanged: the common struggle to get people to take notice, to value the innovation and to agree to help move it along. You will find plenty of examples of that in this book.

David Ellyard
December 2006

Where Was Technology in 1500?

Finding better ways to do things is a quest as old as the human race, beginning with fire, pottery, the wheel and the domestication of plants and animals. From the most ancient times, humans and their ancestors had to provide themselves with shelter, clothing, warmth, food and defence against attack (human and animal). And while much of the technology we use nowadays was developed within the last few centuries in Western Europe and its former colonies, significant pieces of technology have more ancient origins in other regions of our planet, notably Asia.

An immense array of technology originated in the ancient civilisations of China, which continued unabated while the lights were out in Europe following the fall of Rome, though the pace of innovation slackened markedly just as it was picking up in Europe. Significant Chinese inventions that came overland or by sea to Europe—aided by exploration, trade or conquest—included paper, the magnetic compass, sandpaper, water-driven clocks, wallpaper, earthquake-detecting apparatus, noodles, paper money, the toothbrush and playing cards.

What significant inventions took place in the 500 years before 1500 in Europe as it emerged from the Dark Ages? Not a lot, at least in the first few centuries. Most new ideas (or ideas long forgotten) came through contact with Islamic civilisations of the Middle East and with China.

The Eleventh Century

At the beginning of the eleventh century the Chinese perfected the formula for gunpowder, and soon after, printing with movable type, but it was centuries before those reached Europe. The Arab scholar Alhazen knew about lenses, but spectacles were not used in Europe until the thirteenth century. Omar Khayyam's eleventh-century calendar was much more accurate than any used in Europe for many years. However, Islamic science and technology would soon be in decline, suppressed by strict religious teaching that opposed progress.

Not every new idea was imported. The particular requirements of Western music demanded some way of writing it down, rather than each new set of players or singers having to learn it by heart. Around 1036, Italian Guido da Arezzo devised a notation much as we have it today, with the sounds marked as dots on a set of horizontal lines (four to begin with), as well as the do-re-mi system for memorising.

The Twelfth Century

The Dutch led the way in reclaiming flooded land with dams, dykes, canals and pumps driven by windmills; regions in Germany and elsewhere sought their skills and technology. The first mills to make paper appeared in Spain, built by Moors who had carried the innovation west, though most writing was still done on parchment.

Late in the century, the grand new architectural style known as Gothic (initially as a criticism) began to take hold. The Crusaders had brought the concept to Europe from the Middle East. Masons began to experiment with 'flying buttresses' to support ever-higher roofs and spires, raised to the glory of God; these remain impressive achievements today. Domestic housing was more modest, though increasingly windows were filled with glass, chimneys vented the smoke from hearths, and tiles replaced thatch in large cities like London to reduce the risk of fire.

The Thirteenth Century

The magnetic compass, of such value to navigators, was improved (perhaps by Peter Peregrinus, who described it), with the needle on a pivot around which it could swing freely. Previously the needle had been floated on a mat of straw, which was not much good if the sea was rough. European ships began to use the sternpost rudder in place of a sweep oar to control their course. In time, this Chinese innovation would enable much larger ships to be built. Ships also began to sail in convoy to provide greater mutual protection.

The spinning wheel, invented in India, came into use in Europe, significantly speeding the production of fibres for clothing from wool and flax, and foretelling the later large-scale mechanisation of textile production. The production of silk (another import from China, where the silk worm had been domesticated thousands of years earlier) was speeded by mechanical devices that wound silk onto reels, invented probably in Bologna around 1270.

One of the most important imports was the system of numbers we use today (variously called Indian or Arabic), introduced by the Italian Leonardo Fibonacci. This replaced the clumsy Roman numerals and made possible the calculations that would increasingly back up science and technology. In an early commercial use, accountants in Italy invented double-entry bookkeeping around 1270.

English natural philosopher Roger Bacon (who already knew the secret of gunpowder), wrote of the 'magnifying glass' and by the end of the century lenses in frames formed the first spectacles in Europe (again China seems to have had them first). Brandy was first distilled in France around 1300 (chemical distillation apparatus was another legacy from the Middle East).

The Fourteenth Century

Gunpowder arrived on the battlefields of Europe. Invented in China and used in rockets and firearms against the Mongols, it soon found employment in cannons (at the Battle of Crecy in 1346, among other conflicts), small arms, with lead or iron shot fired from iron tubes held against the shoulder (the Chinese had used bamboo), and in rockets. It was a century or more before small arms replaced the longbow and the crossbow, lacking their range, accuracy and rate of fire.

Though sundials were still the most common timepieces, clocks began to chime the hours and quarters from church towers and royal palaces in England and elsewhere (the word 'clock' came from the French for 'bell'). Not until much later did such clocks have faces to show the time.

The Dutch devised 'locks' for canals that enabled a boat to go up or down the side of a hill. Also invented were the form of navigation chart known as a *portolano*, and blast furnaces with water-powered bellows for the production of iron. Sawmills, used in Roman times, were reinvented.

The Fifteenth Century

The pace of invention (and discovery) quickened, spurred by the Renaissance, the many competing courts that sponsored able minds (like Leonardo da Vinci last in the century) and rising prosperity.

This was the century of printing, at least in Europe. Printing with movable type, able to be rearranged to form words and sentences, rather than printing from wooden or metal sheets had been known for centuries in China and Korea. German goldsmith Johann Gutenberg is usually credited as the first printer in Europe, but others may have beaten him by a few years. The innovation spread rapidly across Europe, a major development in itself, and speeding the growth of technology in other fields, enabling news of developments to spread much more quickly.

This century also began the great age of (European) exploration. Portuguese sailors and later those of Spain and other nations voyaged further and further along coastlines and later across open oceans. They were initially looking for routes to the Spice Islands in the East (the overland routes now being blocked by the Saracens), and discovered much unknown territory along the way, such as the unsuspected Americas. These voyages drew on new methods of shipbuilding such as the 'carrack', with exterior planking flush rather than overlapping, and navigation was aided by better mapmaking and the early quadrant used to find latitude.

Musical instruments with keyboards first appeared: the clavichord, with metal bars that struck strings, and the air-powered organ. Instrumental music could now become more complex, with the fingers of both hands employed. The new oil paints, pioneered in Flanders, gave new life and depth to portraits and religious works, largely replacing the 'frescos' painted in wet plaster.

Crafts advanced. Glassmakers found that by adding what we now call manganese to the melt they could remove all colour. The first primitive lathes to turn wooden objects came into use, probably in Nuremberg, driven by a cord moved back and forth by a bow. The carpenter's brace for drilling holes was another innovation of the time.

Medicine remained primitive, though mercury was now used to treat venereal diseases and gold had become the preferred material for filling teeth, in place of the wax and gum used in medieval times. Dentistry (like surgery) was still mostly done by barbers.

Close to 1500, the lives of the people of Europe were constrained by the technologies at their command. They travelled on foot or horseback, unless they were rich or powerful enough to be carried by sedan-chair. Heavy loads went by cart, aided by the horse-collar and shoulder harness, a genuine European invention some 600 years old. There was no faster way to send a written message than to give it to a man on a horse. City streets were usually unpaved, dusty when dry, muddy when wet; country roads were often merely tracks. Water travel was powered by oars or relied on the vagaries of the wind. Nobody travelled by air, except in their imagination.

The only sources of energy to do work were human or animal muscles, though moving water or the wind could be harnessed to specific tasks, such as grinding grain or pumping water. Rich footwear was made from leather, poorer footwear from wood. Fibres for clothing were mostly wool and linen, with spinning and weaving still done by hand and therefore slow, even with the aid of the spinning wheel and the loom. Cotton, later to become so valuable and versatile, was little known in Europe, since its origins were far away in India.

Save for candles, oil lamps and flares, it was dark when the Sun was down and the Moon not up. Coal was only just coming into use, so cooking and home heating depended on wood. Preserving food still relied on ancient techniques such salting, drying, smoking or pickling (or chilling if you had access to ice). Water being commonly unsafe to drink in major centres of population, and new beverages such as tea and coffee yet to arrive from distant lands, most people drank ale with their meal. Soap was a luxury, and not very effective when available, so cleanliness had a low priority; the link between dirt and disease was not understood for another 300 years.

The only explosive—military or civil—was gunpowder; that would be true until the nineteenth century. The dominant metal was iron, especially for weapons and armour, but it was too expensive to allow a farmer to have an iron plough. Its crude production from iron ore, using charcoal from the rapidly disappearing forests, made it brittle, unless extensively reworked by heating and hammering, which again added to the cost. The other metals known were the precious gold and silver, the liquid mercury, copper and tin, which when alloyed made bronze, and the soft unreactive lead, much used in plumbing (from *plumbum,* the Latin word for 'lead'). Zinc was still unknown, so there was as yet no brass.

The practice of medicine was constrained by an almost total lack of the essential knowledge of what was needed to preserve and restore health, mixed with erroneous philosophies that promoted hazardous therapies like bloodletting. External symptoms were the only clue as to what was happening within the body. There was no stethoscope, no clinical thermometer. Lacking the 'germ theory' of disease, and the antiseptic techniques that would ultimately flow from that, doctors were commonly unable to control infection. Anaesthesia for surgical procedures was possible only with the aid of narcotics or alcohol, and surgery generally was rudimentary and risky.

Every one of these varied aspects of life would be transformed in the 500 years to come.

What Time Is It?

Peter Henlein

The first mechanical clocks, dating back to the thirteenth century, were large clunky pieces of machinery that adorned town hall steeples and church towers and were driven by big falling weights. It was hard to imagine a timepiece small enough to be carried.

Around 1500, a locksmith from the medieval city of Nuremberg in modern southern Germany came up with a new power source for clocks. Others may have had the same idea, but Peter Henlein is the one credited with making it work. Henlein took a strip of thin iron sheet, and coiled it up into a spring. Energy was stored in the spring as it was tightened ('wound up') and released bit by bit as the spring relaxed. Machines (such as model trains) powered by such coiled springs are still dubbed 'clockwork'.

The clock could now be shrunk in size, set on a shelf or on a table, or even (and this was Henlein's greatest achievement around 1510) strung on a cord around the neck—portable if not yet a 'pocket watch'.

These early watches were the size and shape of a turnip, though increasingly they were made oval in shape and gained the name 'Nuremberg Eggs'. Often decorated with gems, they became symbols of wealth and fashion. But as timekeepers they were poor, needing careful handling. The crude handmade iron mechanism was stiff, needing a very strong mainspring to make time pass at all. And as the spring 'ran down' the clock ran slower, and time began to drag.

Fixing that problem turned the early watch from a fashion item into a passably useful device. Around 1525, Jacob Zech, a Swiss-trained mechanic living in Prague, conceived the 'fusee'. A cord wrapped around a conical pulley connected the mainspring to the rest of the works, so that the spring maintained a constant pressure and the watch kept much better time.

By around 1600, watches were becoming commonplace. Malvolio in Shakespeare's *Twelfth Night* talks of his with pride. Brass, and later steel, replaced iron in the works, screws supplanted rivets, a glass face protected the 'hands', which now often recorded minutes as well as hours. Such watches were more than ornamental, but still far short of acceptable accuracy. Some way was needed to accurately measure the passing particles of time, count and record them. That took more than a century (p. 29).

A Most Ingenious Mind

Leonardo da Vinci

Few people would dispute that the Italian Leonardo da Vinci possessed one of the most brilliant and imaginative minds in our history, perhaps *the* most. Hardly any field of endeavour in his day escaped his influence: painting, sculpture, literature, music, philosophy, architecture, engineering and science.

He was also a great inventor, at least on paper. He left behind notebooks crowded with diagrams and descriptions of an incredible array of machines and devices for almost every conceivable purpose: clocks, looms to weave textiles, parachutes, mobile cranes, watercraft, weapons of all kinds, underwater breathing apparatus, even flying machines.

How many of these had a great impact on industry or society in his day? The best answer is 'not many'. His canal locks and horizontal waterwheels were good designs; his technique for precisely cutting teeth in gear wheels was influential. But most of the rest remained ideas and were never built. Many could not be made at the time for lack of suitable materials and manufacturing methods. Others proved impractical when attempts were made later to build them. Some did prove workable, if rather out on date.

His flying machines were perhaps the most memorable—a variety of mechanisms to simulate the actions of birds and lift a man into the air. It seems none of them would have worked, though some were similar to today's hang-gliders and ultra-light aircraft. His visionary 'helicopter' (not his term), designed late in life, was too heavy to lift both itself and the man carried inside and also turn the helical screw providing lift. The machine had no tail rotor, so it's likely the whole machine would have turned, rather than just the lifting screw, leaving it on the ground.

His inventions were probably more useful, and more used, in war than in peace. Renaissance Europe was a time of turmoil and conflict. Like other 'natural philosophers', da Vinci was required to devote thought to new methods of attack and defence. Some of his innovations were put into service, including by da Vinci's patron the Duke of Milan. The crossbow was the main weapon on the battlefield: da Vinci developed one that fired several arrows at once, as well as a giant crossbow with greatly increased power. Other war machines included a cannon with three barrels for more rapid fire, and an armoured car with eight men inside turning cranks to drive the wheels.

Da Vinci died in 1513.

Lock, Stock and Barrel

The Technology of Firearms

Sixteenth-century Europe saw many battles and skirmishes between jostling states, so there was much incentive to improve 'firearms'—guns small enough to be carried by a single soldier. The basics were simple. An iron tube, closed at one end (aka the 'barrel'), was mated to a shaped wooden block (the 'stock'), resting on the shoulder or held in the hand. A charge of gunpowder was loaded down the barrel from the open end, followed by some wadding and an iron or lead ball.

Firing of the gun required the charge to be ignited through a small hole ('the touch hole') near the closed end. That required the third member of the 'lock, stock and barrel' trio. Early models from the fourteenth century used the 'matchlock'. Pulling the trigger caused a short piece of slowly burning cord (the 'match') to contact gunpowder in a small pan set above the touch hole. When the pan powder burst alight, flame went through the touch hole to the charge and the gun went off. If it didn't, the soldier was left with an embarrassing 'flash in the pan', and his gun would 'hang fire'.

Matchlocks were certainly better than lowering a lighted match into the touch hole by hand, as the first 'musketeers' did. Infantrymen could have both hands on the gun and eyes on the target. But the glowing match and its distinctive smell could give away the soldier's position at night; rain or mist might put out the match; and a naked flame was always dangerous around gunpowder.

Around 1520 a better idea, the 'wheel-lock', developed, probably in Italy, from the wind-up springs becoming common in clocks (p. 11). When the trigger was pulled, a coiled spring unwound rapidly, causing a piece of flint to strike rough iron, dropping sparks into the firing pan. Wheel-locks were more reliable and easy to use, but also much more expensive. Few foot soldiers could afford them. Matchlocks, with all their problems, stayed in use till around 1720.

Next was the 'flintlock', invented around 1608 in France. Flint striking on iron or steel still made the sparks, but the mechanism was simpler (and cheaper), robust and reliable. The technology had an astonishing longevity. Flintlock pistols and rifles were still in use in the American Civil War, 300 years later. By then new technology was coming (p. 101).

Around 1520 there was another advance. The 'rifling' of a gun barrel—that is, cutting spiral groups inside it to make the ball spin—was first popularised by Gaspard Koller or Kollner, a gunmaker of Vienna, or perhaps by August Kotter of Nuremberg. The spinning ball held its course more truly to the target; shots from a 'rifle' were more accurate and deadly.

New Words and New Languages for Mathematics
Many Ideas from Many Minds

As science and technology began to find their feet through the sixteenth century, the significance of mathematics grew. By the end of the century, Galileo would be ready to declare that the 'story of the universe is written in the language of mathematics'.

Throughout the 1500s, many of the tools and notations we associate with mathematics today first came into use, devised by various mathematicians as they found need, beginning in 1525 with the 'square root' symbol. Negative numbers followed in 1545, the 'equals' sign in 1557, imaginary numbers (like the square root of minus one) in 1572, the use of letters to represent quantities in algebra (such as 'x' for the unknown) in 1591 and the 'decimal point' in 1593.

By 1600, the algebra handed down by the Arabs had been mastered and developed, and the geometry of the Greek authority Euclid rediscovered, along with the other great mathematical works of antiquity. Trigonometry, with its six relations between the sides and angles of a triangle first published by the German Georg Rheticus in 1551, became a separate study. Surveying and astronomy were among the practical activities that benefited.

This process continued though the seventeenth century. A big year was 1631, with the first use of the 'greater than' and 'less than' signs by Englishman Thomas Harriott, and of the modern multiplication sign introduced by his countryman William Oughtred. The year 1654 saw the first use of the symbol for infinity by the Englishman John Wallis; in 1659, Johann Rahn of Switzerland introduced the modern sign for division.

By now, inventive minds were coming up not merely with new ways to write down the existing mathematical languages, but with some quite new languages, capable of expressing new and powerful insights. Seventeenth century French genius Rene Descartes combined geometry and algebra, creating 'coordinate geometry'. In his system, any point in space can be identified by two, three or even more numbers (depending on the dimensions). We still call these sets of numbers 'Cartesian coordinates'.

Half a century later Gottfried Leibniz of Germany and Isaac Newton of England independently created 'calculus' or the 'theory of fluxions'. This powerful way to view the world lets mathematicians deal with changing physical quantities, like speed or position, and to write down equations to describe the motion of objects in time and space. The rivals argued about 'who had invented what' for decades.

The Art and Science of Surveying

Gemma Frisius, Leonard Digges, the Cassinis

The sixteenth century saw a boom in the technology and methods of surveying—precisely locating major features in the landscape and using those to construct maps. Land owners wanted to know how far their estates extended, cities needed accurate measurements for defensive fortifications, new national borders agreed to by treaty had to be firmly established and new lands being discovered (or rediscovered) overseas needed mapping.

Chief among these techniques was the 'plane table'. This was simply a large table on folding legs able to be made perfectly horizontal, with a sighting device that could be used to mark on a chart lying on the table the direction from the observer to a distant object. When a number of such sightings had been made, the table was moved to another spot a known direction away and the sightings taken again.

When the two resulting charts were superimposed, the points at which the sighting lines crossed marked the relative positions of those objects. Initially the plot would give only relative distances, but measuring just one distance, for example, the 'baseline' between the two observing points, would immediately give the distances between all the landmarks plotted.

While this method, known as 'triangulation', probably had many inventors, it was first clearly described in 1533 by the Dutch mathematician, cartographer, physician and instrument maker (a versatility common at the time), Gemma Frisius. A giant in his time, his students included the famous mapmaker Gerard Mercator.

The plane table morphed into early forms of the 'theodolite', beloved of surveyors. The instrument, and its name (whose meaning is uncertain), are usually ascribed to the Englishman Leonard Digges, who described one around 1571. The theodolite could measure vertical as well as horizontal angles and so find heights as well as distances. It often contained a compass to give magnetic directions, and a telescope, once those were invented (p. 23).

Over 100 years elapsed before a large scale survey using triangulation was undertaken, in an effort to correct the many errors in maps of the time. The Italian-born French astronomer Jean Cassini began one of the whole of France, carried on by three generations of his sons and finished in 1740. Early on, King Louis XIV reportedly remarked that the survey (and the amendments to maps resulting) had 'cost him much of his domain'.

The Technology
of Music

Andrea Amati, Adolph Sax and Many Others

The modern symphony orchestra was not 'invented' in a single stroke; it is the result of half a millennium of innovation, beginning with diverse groups of musicians and instruments accompanying dancing or devotion. A symphony orchestra today is built around a core of stringed/bowed instruments—the violin and its larger siblings—and their distinctive sounds, to which an increasing number and variety of blown instruments has been added over the centuries.

String Instruments

The violin and its relatives evolved from the earlier viols, so no one really invented them, but credit as the first great violin-maker usually goes to Andrea Amati from Cremona in northern Italy in the mid-sixteenth century. He applied his lute-making methods to the crude violins of the day, using a mould to ensure precision in construction. He also enlarged the sound box, increasing volume while maintaining sweetness of tone. The oldest known survivor among the Amati violins dates from 1564.

The immediate popularity of Amati's violins began the great Cremona violin-making tradition, including the Stradivarius and Guarneri families, active for nearly 200 years. More innovation came early in the nineteenth century, with the structure of the violin undergoing many changes in search of greater range and volume to suit the virtuosity of the players and increasing size of concert halls, resulting in essentially the modern violin.

Woodwind

The typical orchestral complement of 'woodwinds' also evolved. The (transverse) flute displaced the recorder only from around 1750, joining early oboes and bassoons with their double reeds. The latter had evolved from the more primitive 'bombard' of 200 years earlier.

We can say that the single reed clarinet was 'invented', the first being made late in the sixteenth century by the instrument-maker Denner of Nuremberg. It, too, had its ancestors, including the low-voiced 'chalumeau', but Denner's innovations, such as the speaker key, which allowed the upper range to be reached, made it essentially a new instrument. A century was still to pass before the clarinet found a regular place in orchestras. All the woodwind instruments enjoyed innovation in the nineteenth century and later, with a move to metal construction, changes in the shape of their 'bore' (the tube from the mouth-piece to the bell), and more complex sets of keys, such as the Boehm system, to ease the playing of intricate music.

Brass Instruments

For brass instruments such as cornets, trumpets and horns, the key invention was the valve, introduced around 1815 by the Germans Blumel and Stotzl. Operated by a key or piston, the valves (each instrument had several) enabled the length of the column of vibrating air inside the instrument (which affects the pitch of the note being sounded), to be changed at will, already possible with the trombone, so that all the notes in a scale could be played. The previous technology of 'crooks' to change pitch always left some notes unreachable, just as a bugle has many notes missing.

Through the use of valves, brass instruments gained the capability to play melodies like the strings or woodwinds (or rather regained that ability, which had been possible a century earlier on 'natural instruments' but only through the special technique and extraordinary skill of the players). Finding a bass instrument for the brass section (comparable with the bassoon in the woodwinds) needed further innovation, with various options such as the serpent and the cimbasso being tried before the valved tuba, first made around 1835, took over.

Saxophones

A memorable nineteenth-century name in instrument-making was the Belgian, Adolph Sax, who invented and named a new instrument in the 1840s. The 'saxophone' was a hybrid, made of brass but using the single reed found in a wooden clarinet and with a conical bore like an oboe. Composers were little interested, but saxophones became very popular in French military bands, linking the sound of the woodwinds with that of the brass, moving into American military bands and then into jazz as it emerged, where they found an enduring place. Sax created a family of instruments, with the alto, tenor and baritone in common use, along with the soprano and occasionally the bass.

Sax also created a family of seven brass instruments to produce a uniform sound from highest to lowest. The 'saxhorns' are the mainstay of today's brass bands, going under a variety of names including flugel horn, alto or tenor horn and baritone.

The Tuning Fork

The tuning fork is not really an instrument. It plays only one note, but so reliably sounded that a number of instruments can agree on pitch and play in consonance. Now made of steel, it was invented in 1711 by John Shore, a trumpeter at the court of St James in London. A notable player, he had solo parts written for him by Purcell and Handel.

Something To Write With

Quill Pens, Pencils, Fountain Pens

The ability to record information and ideas in some written form was one of the first steps in civilisation. We progressed from carving on stone and making marks on wet clay to painting symbols on papyrus or parchment and, around AD 700, to the quill pen, made, as the name suggests, from a bird feather.

Nearly 1000 years later the quill pen was still in use everywhere, though paper made from plant fibres (p. 75) had replaced parchment for everyday use. Quill pens needed careful preparation and regular sharpening, using a 'pen knife', and of course regular visits to the inkpot. But the quill 'dip pen' and its metal replacement were well able to deliver the copperplate script that was much admired, and continued to do so through to the mid-nineteenth century.

First to rival the quill was the pencil, invented in 1565 when a thick bed of black material called graphite was found in north-west England. The stuff could make marks on paper, and the name came from the Greek word for 'to write'. Once it was cut into sheets, then into thin rods, and finally encased in wooden holders, we had the first pencils. The name comes from the Latin word *pencillus*, the tiny brush Roman scribes wrote with.

If you made a mistake with the pencil, the error could be easily corrected with the aid of a new material extracted from the sap of South American trees. This could 'rub out' pencil marks, and so was called rubber. For a time, graphite was thought to be a compound containing lead, hence the term lead pencils, but before long it was correctly identified as a form of carbon.

England had a monopoly on making pencils for a long time; no other country had deposits of pure graphite. Later German and French manufacturers made pencils of different hardness by heating powdered graphite mixed with clay.

The quill pen had another emerging rival; the 'fountain' or 'reservoir' pen, which held its own ink supply. These had been around for centuries in crude form, but refinement took a long time. Early fountain pens used quills as nibs, but by the nineteenth century manufacturers were using steel nibs, often gold-plated. Cases and reservoirs were made of rubber, hardened by a newly discovered process (p. 89).

Fountain pens really took off around 1880. Everyone wanted one, and the quill pen finally fell into the dustbin of history.

The Calendar and the Seasons

Pope Gregory XIII

Keeping track of days, months and seasons has always been a matter of concern. Farmers need to know when to plant their crops and when the rains are due, religious leaders when to hold festivals like Easter, government officials when to collect taxes. Everyone has an interest in an accurate and reliable calendar.

Calendars rely on astronomical events and periods: the day, from one noon to the next; the month, the time taken for the Moon to go through its phases; and the year, when the Sun comes back to its same time and place of rising. These units make an untidy mix. There is not an exact number of days in a month, or months in the year. Fitting these pieces together has taxed astronomers and law-makers for millennia.

The Romans, for example, originally thought the year contained exactly 365 days. The year before he was assassinated, and informed by astronomers that there were actually 365¼ days in the year, the leader Julius Caesar took action. He initiated the 'Julian calendar' and the 'leap year', adding an extra day to February every four years to maintain the rhythm.

Unfortunately, a year is 11 minutes short of 365¼ days, equivalent to three days every 400 years. This small error compounded over many centuries to become noticeable. By the end of the sixteenth century, the onset of spring, marked by an astronomical event called the 'vernal equinox', had drifted back by 10 days, pulling the calculated date of Easter back with it. Without some correction, Easter would one day be occurring in February, instead of March or April, which was not acceptable.

The Pope of the time, Gregory XIII, took the advice of his astronomers and changed the rules slightly. In the new 'Gregorian calendar', a 'century year' (like the year 1600) would not be a leap year unless it was divisible by 400. So 1600 and 2000 would be leap years, but 1700, 1800 and 1900 would not.

Countries still loyal to Rome changed to the new system at once, accepting the loss of 10 days in October 1582 needed to bring the calendar back in line with the vernal equinox. Protestant and Orthodox nations were much slower: Britain did not change over till 1752, by which time a couple more days had to be lost, and some people protested at their lives being 'shortened'.

A Machine that Knits

William Lee

We do not know just why William Lee, a clergyman from Nottingham in England, wanted to invent a knitting machine. Reasons offered include: that his wife (or girlfriend) was always knitting and paid no attention to him; that he was weary of the incessant clacking of needles; that he wanted to increase the productivity of hand-knitters and so raise their income; or perhaps to make some money for himself by getting a patent.

Instead of the traditional method, which produced one loop at a time from a pair of hand-held needles, Lee visualised dozens of machine-driven needles producing an entire row of loops, so speeding up knitting immensely. His machine generated a long piece of plain, flat knitting up to 30 centimetres across. It was a remarkable technical achievement for 1589.

That year Lee turned up at the court of Queen Elizabeth with his brother and a patent application. Elizabeth refused him on the grounds that such a machine would take away employment from workers, so increasing destitution in hard economic times (similar words would be repeated many times in future centuries). Certainly many poor people relied upon knitting to supplement their income.

Rebuffed again in 1598, the Lee Brothers crossed the English Channel and found favour at the French court of Henry VI. There, they continued to improve their machine, from an initial coarse eight needles to the inch to 20. Ultimately they could produce fine knitted silk stockings. But their business did not prosper; their patron the King was assassinated and William Lee died penniless in 1610. His brother and their workers returned home.

What happened next is unclear. Some reports say that one of his assistants, a John Ashton, made some crucial improvements and the machines became popular. Others say that the brother founded a workshop in London, making silk stockings for the gentry, or that he worked with a Nottingham businessman to set up the first full-scale knitting factory.

But there is little doubt that the machines or 'stocking frames' became widely used, especially in the Midlands, though not without opposition from handknitters, who feared for their livelihood and broke up many machines. The price of stockings fell to the point where they were no longer a luxury. More importantly, the first step had been taken in the mechanisation of the textile industry, a process that would power the Industrial Revolution in the centuries ahead (p. 42).

Making the Small Seem Larger

Zacharias and Hans Janssen

Around 1590, magnifying glasses were nothing new. For a thousand years or more, we had known that a round piece of glass, thicker in the middle than the edges, could enlarge, say, writing on the page—an aid to weak eyes. Lenses could also focus the light of the Sun, bringing its rays together at one point, and making a 'burning glass'.

No one had done much with this until around the thirteenth century, when smaller, lighter lenses, worn in a frame hooked over the ears and nose, became the first 'spectacles', improving vision. The usual lens, thicker in the middle, and called 'convex', could help people with short sight; and 'concave' lenses, thinner in the middle than the edge, aided the 'longsighted'.

Dutch spectacle-makers were the best in the world in the late sixteenth century. It seems that around 1590 in Middleburg, Zacharias and Hans Janssen were idly playing with some of the lenses in their stock when they found that a concave lens and a convex lens in line made a small close object appear much larger. The two lenses had to be separated by less than the 'focal length' of the convex lens, that is, the distance from the lens to the point at which it focuses light.

Put the two lenses in a tube for easy use and the 'microscope', or more correctly, the 'compound microscope', was born, though the name came much later. It was soon discovered that two convex lenses worked even better. The first to build such a microscope, around 1620, was the 'court inventor' to England's King James I, Dutchman Cornelius Drebble, more famous for reputedly building the first submarine (p. 164).

The later seventeenth century was the great age of exploration with the new microscope. The keen eyes of researchers in Holland, England and Italy saw for the first time 'cells' in plants, bacteria and other 'animacules' in water and wastes, minute blood vessels and other structures in animal tissues, the detailed anatomy of tiny insects. Things previously unseen, and even unsuspected, became visible and knowable.

To be honest, the early two-lens microscopes were not easy to use, and many important discoveries were made, especially by the Dutch, using single lens magnifiers of great power. But the compound microscope won out in the end, and became very powerful and useful, until overtaken by the electron microscope (p. 199).

Flushed with Success

John Harrington, Alexander Cummings

Nothing perhaps better represents an advanced civilisation than the flush toilet. Disposing of human body wastes, and even controlling the smell, has been an eternal challenge, all the more as cities grew more crowded, and evidence was found that contact with excrement could spread disease.

Many people credit nineteenth-century British plumber Thomas Crapper with the invention of the loo as we know it today, not least because of his name (which is really just a coincidence). But key innovations lie much further back. Some archaeologists claim that remains of flush toilets can be found in the ruins of ancient Crete and the cities in the Indus Valley, thousands of years old.

By Tudor times in England, all that refinement had vanished: 'privies', where people might do their business in a bucket, were not nice places to be. The most up-to-date equipment, popular with the rich and powerful, was the 'close-stool', simply a comfortable padded seat above a metal or porcelain container, which of course still had to be emptied.

The man who moved us on was Sir John Harrington, courtier to Queen Elizabeth I and often in trouble for telling risqué jokes. He is reputed to have invented the Ajax flush toilet for the queen in 1596. This 'water closet' did not need emptying. Rushing water from a cistern set above the unit washed excrement and urine into a trap beneath the bowl, then via a drain to some convenient repository, like the castle moat or a nearby stream (that issue would not be addressed for several centuries yet). The invention did not really catch on.

Various designs for valves to separate the bowl from the trap and drain followed, but the key innovation, patented by Englishman Alexander Cummings in 1775, was the 'S bend', brilliantly simple. This curved tube kept an air and water seal between the bowl and the sewer, so preventing any backflow or nasty smell. That's still what we use today.

And what of Thomas Crapper? It seems he did not really invent anything, but rather manufactured S-bend type bowls to someone else's design. He did improve the operation of the cistern and he made the units so well that he received a royal warrant to provide toilets for the monarchy. Fame, and the use of his name as a familiar designation for a WC, followed inevitably.

Making the Far Away Seem Closer

Hans Lippershey

We don't know for sure who invented the telescope. Writers over previous centuries, including Englishmen Roger Bacon and Leonard Digges, had alluded to using mirrors and lenses to make distant objects seem closer. But no one seems to have made them work.

We do know that the first to try to patent a telescope was Dutch spectacle-maker Hans Lippershey in 1608. He had found, perhaps by accident, that looking through a concave and a convex lens in the same line made far-away objects appear larger, provided that the lenses were appropriately spaced . Twenty years earlier, a different arrangement of lenses had created the first microscope (p. 21).

Lippershey did not get his patent, mostly because other people were trying to do the same. News of these 'optick tubes' spread quickly, including to Italy, where the physicist Galileo made one of his own, improving the technology along the way. He demonstrated that the telescope could, for example, detect and identify ships when they were still far off, a benefit in war and commerce.

He was first to turn the telescope to the heavens, initiating a century of stupendous astronomical discoveries by himself and others: mountains and craters on the Moon; spots on the Sun; the visible disks of the planets with their phases, spots, rings and orbiting satellites; and the vast number of stars making up the Milky Way, a first insight into the immensity of the cosmos.

There were adjunct benefits. Copernicus's still-controversial theory that the Earth rotated around the Sun (rather than the opposite) gained new support. Galileo's discovery of the moons of Jupiter, which he initially named after the Medici family (the rulers of Florence) secured him a well-paid job in their court, with time and resources for other studies.

Different combinations of lenses soon made telescopes work better. They became more powerful, but limited by defects in the lenses. Confusing rings and smears of colour contaminated the images. Believing these to be unavoidable using lenses, Englishmen James Gregory and Isaac Newton independently devised 'reflecting telescopes', with a parabolic mirror at the bottom end of the tube to collect and focus the light, rather than a lens at the top end, as in Galileo-style 'refracting telescopes'. Both sorts of telescopes had their supporters.

Telescopes grew in size astonishingly over following centuries, gaining sophistication and power that Galileo and Newton could never have imagined (p. 64).

Mathematics Made Easier

John Napier, William Oughtred and Others

As the deeply religious Scottish genius John Napier put it:

> ...seeing there is nothing that is so troublesome to mathematical practice, or doth more molest and hinder calculators, than the multiplications, divisions, square and cubic extractions of great numbers, I began therefore to consider in my mind what by certain and ready art I might remove those hindrances.

Napier's response to the challenge in 1614 was the system known as logarithms or simply 'logs'. This makes every number a 'power of 10', reducing multiplying to adding, dividing to subtracting. To find a square root, simply divide the logarithm by two. It was a revelation.

Around 1622 the English clergyman and mathematician William Oughtred inscribed logarithms onto two metal disks on the one axle, so they could be aligned for the purposes of calculation. From this he developed, in around 1633, the 'slide rule', a device found in the breast pocket of most scientists for the next 300 years, until the electronic calculator took over in the 1970s.

Long before electronics, clever minds were devising 'calculating machines', in which gearwheels and cranks would replace mental effort in doing sums. The German Wilhelm Schickard was the first with his 'calculating clock' in 1623. This could add and subtract six-digit numbers and devise tables giving the positions of the planets and other astronomical objects.

Others to go down the same path included the pensive French philosopher Blaise Pascal. His 'Pascaline', invented in 1642 when he was only 19, was designed to relieve his father, a clerk, of much tedious calculation. It worked tolerably well, but only Pascal himself could repair it, and the need to cut the gearwheels by hand meant it cost more than the people it replaced. Still, mathematicians began to fear for their jobs.

The German Gottfried Leibniz, the great rival of Isaac Newton, worked on the problem later in the century and found a way to speed up multiplication and division by treating them as a series of additions and subtractions. Any mechanical calculator still working today uses this principle.

For all their ingenious complexity, the machines were crude and unreliable, prone to breakage, hard to repair, a consequence of the ill-formed state of manufacturing at the time. Furthermore, they could not be 'programmed', needing the constant intervention of the operator. Not until the nineteenth century could a mechanical calculator work unsupervised (p. 80).

1626

'The Effecting of All Things Possible'

Francis Bacon

Francis Bacon, the seventeenth-century English philosopher, writer, lawyer, statesman and a major figure in Stuart England, was not an inventor. He was not really an experimenter either, though the story persists that his death at the age of 65 was the result of a chill he caught when he stepped from his carriage on a wintry day to discover whether a chicken could be preserved by being stuffed with snow.

But he was a visionary. His little book *New Atlantis,* published just before his death in 1626, would have been called 'science fiction' had that term been in use. It tells of a journey to an imaginary island in the South Seas. There, an advanced culture is represented by the remarkable 'House of Solomon', which we in our time would think of a major research institute or laboratory, and which stimulated the formation of the Royal Society half a century later.

Late in the book, the travellers meet the head of this house and are told (apparently in Spanish) about its purposes and activities: 'The end of our Foundation is the knowledge of causes, and the secret motions of things, and the enlargement of the bounds of human empire, for the affecting of all things possible'. Translated, this implies that both science and technology were being pursued in the House of Solomon, with newly discovered knowledge about how the world works being directed to useful ends.

Bacon set out some of these targets in an appendix to the book. Taken together they form an outline of a possible program of far-reaching technology development. It is a miscellaneous wish list, not even well ordered, but interesting in that it contains the sort of 'things' that Bacon apparently considered 'possible'; in other words, circumstances and situations able to be affected by the application of knowledge. It is worthwhile to ponder how many of these have been achieved in the 400 years since.

In the area of medicine, Bacon apparently thought the following objectives worth pursuing: prolongation of life, restitution of youth (in some degree), delaying the onset of ageing, increasing strength and activity, mitigation of pain (and better capacity to endure it) and better and gentler medical treatments ('more easy and less loathsome purgings', as he put it), together with 'the curing of diseases counted incurable'. Those ends are still sought today, with notable if varying success.

Some things might come under cosmetic or orthopaedic surgery: 'altering of complexions, including fatness and leanness', 'altering of statures', 'altering of features'; and in a different category, 'increasing and exalting the intellectual parts' (in other words, technology to make people smarter). 'Exaltation of the spirits and putting them in good disposition' might translate today into 'developing anti-depressant drugs'.

Military needs were covered: 'instruments of destruction, as for war and poison'. 'Impressions of the air and the raising of tempests' suggest a capacity to alter the weather. Bacon looked to many processes to be speeded up through research: germination of seeds, purification of liquids, 'putrefaction', extraction of minerals by heating and maturing of crops. Perhaps new species could be made, or one species transplanted into another (as in *The Island of Dr Moreau*).

In industry, Bacon saw scope for synthetic materials, such as paper or glass or fibres for clothing, 'artificial minerals and cements', new sorts of food 'drawn out of substances now not in use', the capacity to make substances much harder or much softer, and major chemical transformations ('turning crude and watery substances into oily and unctuous substances').

Some of the vision is more problematical or less clear. We could handle, and indeed have, 'deceptions of the senses' or 'greater pleasure of the senses'. Most obscurely, Bacon had on his list 'force of the imagination, whether on another body or on the body itself'. Unless he was talking about the placebo effect, this is one of Bacon's targets we are nowhere near hitting.

As Bacon described it, the House of Solomon had an impressive array of facilities ('preparations and instruments'), where investigations in search of all these ends were performed. The staffing was equally diverse: some sailed the world in search of news of experiments, others ploughed through the literature; some devised and performed experiments, others wrote up the results; still others interpreted the outcomes.

But not all the results of these investigations, not all the new technologies resulting, would be made public. Some, if only a few, would be kept secret, even from 'the state', if the toilers in the House of Solomon thought that best. Bacon seems to be indicating that not all new technology represents progress, that there is a distinction between 'all things possible' and 'all things desirable'. The next 400 years would find more than a grain of truth in that.

Measurement Matters

William Gascoigne, Pierre Vernier

In manufacturing today, precise measurement really matters. Mass production using interchangeable parts is possible only because the parts can be measured with precision and so made essentially identical. Two tools to provide such accuracy, the micrometer and the vernier scale, were invented in the seventeenth century, but to meet the needs of astronomers and surveyors. Use in industry came much later.

The Micrometer

Aristocratic William Gascoigne was killed in 1644, aged only 24, during the English Civil War. Keen on astronomy, he had wanted to know how far apart two stars were in the sky. In his 'micrometer', turning a screw caused a hairline marker to advance across the field of view from one star to the other. Noting how far the screw was turned measured their angular separation.

Gascoigne may never have built one of these, having died so young, but he certainly floated the idea, and other astronomers, such as the great architect Christopher Wren, took them up. At that time, precision was much less important in industry than in astronomy, so a century passed before the instrument was used in manufacturing, initially by James Watt (p. 77).

The Vernier Scale

Pierre Vernier's 'vernier scale', a cunning way to increase the accuracy of measurements, is still in use. Vernier was born a quarter century before Gascoigne, in France–Compte (eastern France), then controlled by Spain. War between France and Spain over the territory was a constant threat; Vernier worked as a military engineer, building fortifications. That required surveying; he collaborated with his lawyer/engineer father on a map of his region.

The vernier scale was first described in 1631 in a book dedicated to the son of the Spanish king. This method uses two scales side-by-side. If the main scale was marked in tenths of an inch, the vernier allows the user to visually estimate length to a hundredth of an inch, essentially by noting how the markings on the two scales, constructed slightly differently, line up.

Vernier used his scale firstly (and mostly) for surveying, applying the same principle to measuring angles. With the main scale marked in half degrees, the vernier gave a reading in minutes, units 30 times smaller. Any measurement on a graduated scale, such as air pressure from a barometer (p. 28), could now be made with unprecedented accuracy. Precision measurement would be ready for use in the industrial age when that came.

Can We Use the Power in the Air?

Evangelista Torricelli, Otto von Guericke

Practical people took a long time to realise that air has weight and to make use of the fact. Popular wisdom, following Aristotle, said that air was intrinsically light and always wanted to rise. He had also argued that a 'vacuum', a space empty of air or anything else, was impossible. Not so.

Italian Evangelista Torricelli, secretary and assistant to the ageing Galileo, made the key discoveries in 1643. In a classic experiment, he filled a glass tube, closed at one end, with mercury and inverted it into a basin of mercury. Some of the mercury ran out, but not all. The remaining column of mercury, about 750 millimetres tall, was apparently held in place by the weight of the air pushing down on the mercury in the basin. That same force would support a column of water 10 metres high, which answered an old puzzle faced by the builders of fountains and waterworks. Why could water not be sucked up to a height of more than 10 metres?

Furthermore, there was now a space between the top of the mercury and the top of the tube, which clearly contained no air. Here then was Aristotle's 'impossible' vacuum.

Toricelli's invention was a barometer, a device to weigh air. Modern 'aneroid' barometers need no mercury. Their pressure readings are of vital importance to weather forecasters.

Air pressure is massive, equivalent to 1 kilogram weight on every square centimetre of any surface. After reading about Torricelli's work, the German engineer Otto von Guericke, later mayor of Magdeburg, staged a striking demonstration in front of the German Emperor in 1854. He had devised a pump to pull air from an enclosed space, similar to a pump used to raise water but made as airtight as possible. He took the air from a sphere of metal, made in two pieces. With a partial vacuum inside, the unbalanced air pressure outside was so great that two teams of eight horses could not pull the halves of the 'Magdeburg hemispheres' apart.

Von Guericke wondered if such unbalanced air pressure could do useful work. In 1672, he removed most of the air from one end of a cylinder fitted with a tightly fitting piston. The combined strength of 20 men could not stop the external air pressure from pushing the piston into the partial vacuum. Interesting, but such an 'atmospheric engine' was not yet practical. How could he make the vacuum without an air pump and much labour? There was a way (p. 37).

Keeping Better Time

Galileo Galilei, Christiaan Huygens

The sort of clocks known to the multi-talented Italian physicist Galileo Galilei could lose or gain many minutes in a day despite 150 years of development (p. 11). Galileo found the way forward by chance. In 1581, when only 17 and a medical student at the University in Pisa, he was standing in the cathedral there (so the story goes) watching as the great lamp in front of the sanctuary swung from side to side.

As the swings got smaller, Galileo timed each one by counting the beats of his pulse. Each took the same time, whatever its size. Only the length of the chain appeared to affect the 'period' of the pendulum. A swinging pendulum a metre long marks off equal seconds. If these units could be tallied to mark the passage of minutes, hours and days, accuracy in timekeeping was assured. Such a clock should not run fast or slow at all if kept in the same location and the same conditions.

Galileo himself saw the possibility of a 'pendulum clock' and drew a diagram towards the end of his life, but never built it. That honour belongs to the Dutch genius Christiaan Huygens, famous in astronomy and mathematics as well as clock-building. In 1656 he built a clock far more accurate and reliable than any built before. Each precise tick of the pendulum was passed through the mechanism known as an escapement (Huygens invented a new type of that too), to record the passage of time on the face of the clock. Scientists were probably the first to use such clocks; precise timing was becoming important in physics and astronomy. It mattered less in everyday life at the time, although Huygens patented accurate pocket watches as well.

Huygens had another vision, a solution to the vexed problem of finding the position of ships out of sight of land. If such a clock could be taken to sea, carrying within it for weeks at a time an accurate memory of the time at the home port, the captain or navigator could compare that time with local time as determined by the Sun and so work out how far east or west the ship had sailed. Huygens strove to build such a clock but it eluded him. Success in that, and the problem of 'the longitude', came in the next century and from across the Channel (p. 50).

Give Blood, Give Life?

Jean Denys, James Blundell

Blood is the stuff of life: sustained loss leads inevitably to death. Stories of blood transfusions go way back, although in the early cases blood extracted from one individual was usually drunk in a tonic by another, since the circulation of the blood was not understood. The first authenticated report of a blood transfusion is from 1665. English doctor John Wilkes took about 60 millilitres of blood from one dog using a syringe and injected it into another, with no harmful effects.

Details of the first animal-to-human transfusion are confused, but the year was probably 1667 and the doctor Frenchman Jean Denys. One account says sheep's blood was transfused into a 15-year-old boy, who later died, causing Denys to be accused of murder. In another version, the newly wed patient was not sick, just prone to bouts of drinking and debauchery. Denys thought he could cure him of this by transfusing blood from a calf, a gentle animal, free from vice. The patient complained of kidney pain and passed dark urine and, although he survived, declined further treatment.

In 1818 British obstetrician James Blundell took the next step, human-to-human transfusion. Concerned by the death of women from haemorrhage after childbirth, he took about 150 millilitres of blood from the arm of the husband of such a patient with a syringe and injected it into her. Over the next few years, he repeated this procedure a number of times, with half of the patients surviving, and developed special instruments for the task.

Deaths from transfusions were not uncommon in those early days, due (we understand now) to the presence of different antibodies in the donor's and recipient's blood causing blood clots or other damage. Animal to human transfusions ceased in the late nineteenth century, but a real answer came only in 1901 with the discovery of blood groups (A, B, AB, O) and the inherent incompatibility of some blood types. Ways were also found to store blood for a time, using anticoagulants and refrigeration, so that transfusions did not need to be made direct, and the way was open for blood banks such as we see today.

Many of the early transfusions succeeded through luck. The distribution of blood types among Europeans allows transfusions with randomly chosen blood to succeed in more than 60 per cent of cases.

The Bubbles in the Wine

Dom Pérignon

When the French monk and cellar master Dom Pérignon first tasted his new brew, a wine with deliberately induced bubbles, he is reputed to have so enjoyed the experience that he cried out to his fellow monks 'I have been drinking the stars'.

Not quite so. Initially at least, the good monk was concerned to keep the bubbles *out* of the wine, since they were unwanted and surely a sign of poor winemaking practice, at least in France. Fizzy wines were popular in England; molasses or sugar was often added, particularly to wines imported from the Champagne region of France, to restart fermentation and fill the liquid with fresh bubbles of carbon dioxide. In the initial fermentation, in which yeast acts on sugar to produce carbon dioxide and alcohol, the gas escapes.

The English had two technological advantages in dealing with 'sparkling' wines. They sealed their wine bottles with cork, which trapped the gas inside. (The French were still using wooden stoppers wrapped in hemp.) And they used stronger glass in their bottles, so they were less likely to burst.

It was an accident of climate that caused bubbles to first show up naturally in wine from Champagne, where Dom Pérignon's abbey vineyard of Hautvilliers lay. For some decades, the weather had turned very cold, shortening the growing season and ending fermentation early. In the spring, the warmer weather caused fermentation to start again in the bottled wine, so producing the bubbles.

Initially thought undrinkable by the nobility, 'champagne' later became very popular in royal courts, especially in London. Dom Pérignon, who had been sent into the region to try and save it from the plague of bubbles, now needed to make more bubbles, not fewer. By trial and error, he developed an early version of the 'méthode Champenoise', putting extra sugar, and later extra yeast, into the bottled wine.

He started using corks and stronger bottles, blending wine from various parts of his vineyard to get more consistent quality. He also made white wine from red grapes, which no one had done before. Soon everyone was copying his methods.

And what is the appeal of bubbly wine? It seems that the bursting of the bubbles enhances the release of flavours from the wine, making it a treat for the nose as well as the tongue. Thank you, Dom Pérignon.

New Forms of Glass

George Ravenscroft and Others

Glass is an ancient invention, much refined by generations of trial and error. Once expensive, it is now cheap and versatile, with some drawbacks, such as its potential to break into jagged fragments. Scientifically, glass involves a reaction between the mineral silica (silicon dioxide, the main constituent of sand) and various other minerals in small amounts such as potash, soda or lime, melted together at temperatures of nearly 2000°C and allowed to harden. Progress in glassmaking has paralleled mechanisation in other industries. For example, mechanical bellows for routine or low-value products like bottles replaced the skilled human glassblower long ago.

The first real advance on ancient glass technology came in 1676, when Englishman George Ravenscroft added some lead oxide to the glass mix. The result was remarkable. The emerging glass was relatively soft to cut, heavy in the hand, brilliantly clear with flashes of light and colour due to its higher 'refractive index', and rang musically when struck. The product was known as 'flint glass', since a few years earlier an improved glass had been produced by adding crushed flint (a mineral composed almost entirely of silica) to the mix. We nowadays call it lead glass or 'crystal'.

Glass now has many forms, mostly developed in the twentieth century. Glass blocks or 'bricks', able to replace ordinary bricks in a building, and 'fibreglass', masses of glass fibres thinner than a human hair, both came into use in 1931. The latter is as resistant to chemical attack and weathering as ordinary glass, and is used for heat insulation. Implanted in a matrix it forms a 'composite' in all sorts of building applications including boats and airplane components. Fibreglass composites combine great strength with light weight. Glass 'optic fibres' are now of critical importance in communications.

Objects made from borosilicate glass (glass containing boric acid) such as Pyrex® were found to be much more resistant to shock and heat than ordinary ('soda') glass. Glass carefully treated by heating ('tempered') becomes much stronger and when broken forms into rounded fragments rather than hazardous shards ('safety glass'). Since 1957, 'glass ceramics' (for example CorningWare®) have been created with millions of tiny crystals, where ordinary glass has none. They are therefore tougher and more heat-resistant, and able, for example, to go directly from freezer to microwave to table; so they find a place in many kitchens. Window glass has had its own challenges (p. 237).

Robert Hooke's Joint

Robert Hooke

Englishman Robert Hooke had so many achievements in so many fields he was compared by admirers with Leonardo da Vinci. He could write well, draw and paint. He invented all sorts of devices. He was deeply involved in the great scientific discoveries and disputes of his day, such as the use of the microscope. He surveyed and helped rebuild London after the devastation of the Great Fire (1667); and he played a leading role in the founding of the Royal Society of London. He is little remembered today largely because his scientific world was dominated by the even greater Isaac Newton, a lifelong rival.

Hooke was short, ugly (a victim of smallpox), orphaned young, argumentative, gregarious and driven by fierce energy from one project to the next, usually leaving them incomplete. But some of them he finished. The 'universal joint' was one.

Hooke did not invent the joint: Italian mathematician Girolamo Cardano had proposed it a century before, and he may have got the idea from the old Chinese technology we now call 'gimbals'. Three concentric rings are interconnected so that whatever is supported in the centre of the three, say a candle or a compass, always remains upright no matter how the surroundings move. Around 1545 Cardano had theorised that the same idea could transfer rotary motion through an angled joint, but he never built one.

Hooke did, producing a working model in 1676. He needed something like it for an instrument to track the Sun across the sky. He connected two rods by a joint rather like two stirrups back to back, so that twisting one rod turned the other, no matter what the angle between them. In this way, a rotating axle could be made to 'turn a corner'. After Hooke, not much use was found for the 'Hooke joint'; we had to await the age of machinery, and particularly the invention of the automobile, for its value to be really appreciated.

A complete list of Hooke's inventions would take pages. It would include: a greatly improved air pump (p. 28); a way to accurately cut the cogs on a gear wheel; the iris diaphragm to control the passage of light (later a vital component of cameras); the much more reliable 'anchor escapement' for watches (p. 29); a 'weather clock' to record changing air pressure and temperature on a revolving drum. All these and more he proposed; some he built.

Forgotten Pioneer of Steam

Denis Papin

Denis Papin is little remembered nowadays, but this inventive seventeenth-century Frenchman has at least one legacy still active, the pressure cooker, or as he called it, the 'steam digester'. This was (and still is) a pot with a tightly fitting lid, which allowed internal steam pressure to build up when the water-filled pot was set boiling on the stove.

Water boils at a higher temperature as the pressure rises, and the combination of higher temperature and steam pressure was very effective in cooking tough meat and even softening bones. If the pressure rose too high, a valve on top opened to let out the excess steam—the first 'safety valve' as later used on steam engines.

Papin had originally trained as a doctor but turned to science and invention early on. In Paris he helped the eminent Dutchman Christiaan Huygens, builder of the first pendulum clock (p. 29), with investigations into air pressure, but also with unsuccessful attempts to harness the force of gunpowder, the only explosive known at the time. Rapidly expanding hot gases from the burning powder were supposed to drive a piston in a cylinder, but the iron of the day could barely withstand the explosion, and the experiments went nowhere.

Papin was a Huguenot (a French Protestant) and like many others suffered from the increasingly hostile attitude of the Catholic King Louis XIV, which led in 1685 to the withdrawal of their right to practise their religion in public. In consequence, he spent much of his life in exile (voluntary or enforced) in Germany and England.

Papin was in London through his late twenties and early thirties, working with the Irish aristocrat Robert Boyle on the properties of gases. With Boyle's support he became not only a Fellow of the Royal Society of London but also gained some employment as the 'Assistant Curator of Experiments'; Robert Hooke was the Curator. It was in front of the Royal Society that he first demonstrated his steam digester in 1679.

Papin was one of the many godfathers of the steam engine, if not actually a parent. Like any number of researchers at the time, he dreamt of using the force of steam, perhaps combined with air pressure, to drive a piston back and forth in a cylinder and so do some useful work, particularly pumping water out of mines or raising water to supply fountains.

In his design, a vertical cylinder closed at one end had a tightly fitting piston with water trapped below it. The water was heated by an external fire, boiling into steam, which then drove up the piston. When the exterior of the cylinder was cooled with water, the steam condensed, leaving a partial vacuum, and unbalanced air pressure (helped by gravity) drove the piston down again. The cycle was then repeated. The movements of the piston could drive a pump.

So the tube was combined boiler, working cylinder and condenser, not very practical or efficient. A better and more successful design came at almost the same time from Englishman Thomas Newcomen (p. 37); this had a separate boiler to supply the steam. And even better results came when Scotsman James Watt added an external condenser (p. 51).

It is not certain that Papin ever built such a machine, though he did work on the idea for years. We know of it mostly from a long series of letters he wrote to the great German mathematician Gottfried Leibniz, with whom he had worked. In those letters he also wrote of a small carriage he had constructed driven by the force of steam. If this was actually ever built, it would have preceded Nicolas Cugnot's 'steam dray' by half a century (p. 52).

He described a steam-powered gun or slingshot (a 'ballista') of such power that it would bring the military ambitions of France towards the German states, where he had a university post at the time, to an end. There is also a story that he put a small steam engine in a model boat, but boatmen watching him at work were so incensed by what they saw as a threat to their livelihood that they destroyed it. On another occasion, an ingenious boat in which paddle-wheels replaced oars was confiscated for much the same reason.

Clearly Papin did not make any serious money from these inventions. By 1707 he was heading back to London with his family and not much else. There he had been previously well received, and had hopes of financial support to continue his research. But times had changed. The scientific colossus Isaac Newton, a great rival of Leibniz (and of Huygens), was now in command of the Royal Society as its president. Papin received no help and died in 1712 in obscurity and (presumably) poverty.

Living and Working Beneath the Sea

Edmond Halley, William Beebe, the Piccards

Englishman Edmond Halley is best known for the comet named after him. But with immense talents and diverse interests, he made major contributions in physics, mathematics, astronomy, technology, navigation...and undersea salvage.

Halley did not invent the diving bell, a type of 'metal barrel lowered into the water upside down', as it had been described by his predecessor, philosopher Francis Bacon. One had been used to recover cannons from the sunken Swedish warship *Vasa* in 1628. However, Halley made some major improvements to the design and operation of the bell, and willingly went undersea himself to test the device out.

His wooden diving bell was weighted with lead, with glass windows, a platform to stand on and benches for seating. It hung by a rope from a spar on its parent ship. To keep the air in the bell fresh, Halley released stale air from a valve at the top of the bell and sent down new air in lead-weighted barrels.

Men could work 20 metres under water for several hours, carrying out salvage operations on underwater shipwrecks, recovering cargo and cannons. Engravings at the time show men working outside the bell connected to the air supply inside via a 'cap of maintenance', a sealed helmet with an air hose attached. Halley ran a profitable business doing such work, helping restore the family fortunes after the ravages of the Great Fire.

Even at that depth, external pressure would have forced water in to fill two-thirds of the bell. That obviously put a limit on how deep a bell could go. The water could be kept out in greater depths by pumping in more air and increasing the pressure, but putting the occupants at risk of 'compression sickness', as it was later known.

To dive deeper required a closed vessel, able to withstand the water pressure, which increased steadily with depth. Of course, the occupants would be unable to leave and return freely, and so would be limited to looking. Not till 1934 did American explorer William Beebe dive 500 metres into the Mediterranean in his 'bathysphere', suspended, like Halley's bell, on a cable.

To go still further down, the vessel needed to float untethered. In 1953 Swiss explorers Auguste and Jacques Piccard descended more than 3 kilometres in their 'bathyscape' *Trieste*. In 1960 the *Trieste* reached the deepest part of the ocean floor, the 11 000-metre deep Marianas Trench near Guam.

'The Miners' Friend'

Thomas Savery, Thomas Newcomen

The world's first practical steam engine did not drive any machinery; it pumped water from coal and tin mines. Driving ever deeper in search of deposits, miners reached the water table, a mostly constant level of water permeating low-lying rocks. Keeping mines adequately dry was a great labour, cutting profits severely. A mine might need 500 horses to drive pumps. The inventor of the steam-driven pump, Englishman Thomas Savery, called his invention 'The Miners' Friend'. His pamphlet, *An Engine to Raise Water by Fire*, appeared in 1702 but one of his pumps had been running successfully since 1698.

The operating principle was quite simple. Three pipes fitted with one-way valves were connected to a large metal cylinder. The cylinder was filled with steam through one pipe connected to a boiler. A spray of cold water condensed the steam, creating a partial vacuum. Outside air pressure then pushed water up a second pipe from the mine to fill the cylinder, much like water sucked up through a straw. A steam blast drove the water out again, upwards through the third pipe, leaving the cylinder full of steam. The process was then repeated. Frenchman Denis Papin (p. 34) had devised something similar, though his cylinder was also the boiler, which made it less efficient.

It worked but not very well. The height to which the water could be raised depended on the pressure of the steam. That was limited by the strength of soldered joints in the boiler and cylinder. A 20-metre vertical lift was about the limit. More than that needed several pumps in series.

Savery's model was overtaken in 1712 by a design from his business partner Thomas Newcomen. This separated the pump from the engine. The cylinder, open at the top, now contained a piston. Steam pressure pushed the piston up. When the steam was condensed, unbalanced air pressure and gravity forced it down again. Here was an 'atmospheric engine', such as Otto von Guericke (p. 28) had proposed. The movements of the piston rocked a beam to drive the pump.

Newcomen's engine was soon driving pumps all over Britain and Europe. At the time of his death, a hundred Newcomen engines were in service. Though they were cheaper than horses, users had to tolerate their slowness and wastefulness for more than 50 years (less than 1 per cent of the energy in the coal used to raise the steam did any useful work). But ultimately, better technology did arrive, thanks to James Watt (p. 51).

Seeds in the Ground

Jethro Tull

Around the turn of the eighteenth century, crops like wheat or barley or oats were usually planted, wastefully, by hand. Seeds were broadcast, much as had been done in biblical times, flung out in handfuls onto ploughed land. The seeds did not spread evenly, too few here, too many there. They were not automatically covered by soil, so many would fail to germinate; birds would eat others.

The man usually credited with finding a better way, Englishman Jethro Tull, trained as a lawyer, but could not afford to start a practice. So he took to managing the family farm in fertile Berkshire (he named it Prosperity Farm, perhaps more in anticipation than actuality). An early invention, or perhaps reinvention, around 1701, was his 'seed drill', a machine on wheels pulled behind a horse. This planted three parallel rows of seeds in furrows, at regular intervals and the right depth, covering the seeds up as it went, According to some reports, he made his first working model using parts from an old pipe organ.

This was obviously a good idea, reducing seed wastage and increasing productivity (up to eight times, Tull claimed). Since the growing crops were in neat rows, it was much easier to keep the ground between them free of weeds, using a horse-drawn hoe, another of his inventions. But the seed drill was slow to take on, especially in England; it was more popular in France and the American colonies. Farmers were a conservative lot, and Tull's ideas seemed radical.

Some of his ideas were wrong. He did not like using animal manure to keep fields fertile, since it often contained the seeds of weeds, and recommended repeatedly breaking up the soil into tiny particles to 'release the nutrients'. This drawn-out (and ineffective) process often delayed planting and made late-growing crops vulnerable to diseases. His advocacy here may have made others slow to take up his good ideas, such as the seed drill and improvements to the plough, and the use of horses instead of oxen.

Tull was poor, ignored and even ridiculed most of his life, but his inventions, popularised by his writings, won through in the long run. The increased capacity of the land to produce food and feed people was one factor that made possible the future growth of the big industrial towns.

The Secret of Porcelain

Johann Bottger

Pottery goes far beyond technology and functionality into the realms of artistic creation, elaborate forms, brilliant decorative colours and patterns, gilding and other refinements combining to generate works of art for everyday use. In eighteenth-century Europe, the pottery most sought after for cups, plates and other crockery was white, semi-translucent, durable but delicate in appearance, called 'china' from its place of origin, or 'porcelain' since it resembled a certain type of seashell.

Chinese artisans had perfected the process over many hundreds of years, using a unique mix of pure white kaolin clay and the mineral petuntse, found only in China. When baked in the oven, petuntse melts and fuses with the kaolin, producing the unique qualities of 'hard-paste' or 'true' porcelain. The Chinese guarded the secret for centuries.

Early explorers and traders returned home with samples of 'china' (among other wonders). With the growing popularity of new beverages from the Americas and the East, such as tea, coffee and chocolate, demand for porcelain cups and saucers grew rapidly among the well-to-do. European manufacturers tried to emulate china but could not, lacking the right minerals.

Around 1575 experiments in Italy led to 'artificial' or 'soft paste' porcelain, made by firing mixtures of kaolin and local glassy materials. It remained slightly porous, but had a creamy colour that some preferred to the bright white of true porcelain. Over coming centuries, the manufacture of soft-paste porcelain spread across Europe, with still-famous factories founded at Sèvres and Limoges, near deposits of high-quality kaolin, and also in England, Germany and Italy.

But it was not yet the 'real thing'. In soft paste, the outer 'glaze' is distinguishable from the inner 'body' of the piece; in 'hard paste', they cannot be separated. The secret was not known in Europe until 1709. German chemist Johann Bottger realised that the Chinese were using a mixture of minerals. He methodically tested samples from all over Saxony to find what he needed. His factory at Meissen (not far from Dresden) was famous for decades; 'Dresden china' remains well known. Bottger died aged only 37.

Around 1750, a new form of porcelain, 'bone china', was invented by Josiah Spode. Strongly heated ('calcined') and crushed ox bones were added to the soft-paste mix. The result was harder than soft paste, though not as durable as true porcelain, with an appealing translucence. Bone china became particularly popular in England.

Keyboards that Play Loud and Soft

Bartolomeo Cristofori

Making music on 'keyboard' instruments goes back a long way, at least to the invention of the organ, perhaps 1000 years ago. Pressing keys on one of these let air into pipes of various lengths to make the different notes.

Not quite as ancient are the harpsichord, and the older related virginal and spinet. In these, the keys caused 'plectra', made from feathers or points of leather, to pluck strings, one for each note. The drawback with these instruments for more expressive music-making is that they are not 'sensitive to touch'. No matter how you press or bang the keys, the resulting notes have the same volume.

Bartolomeo Cristofori was a harpsichord maker from Florence. In 1709 he published the plans for a new type of keyboard instrument which could be played loud and soft: 'forte' and 'piano' in Italian, hence 'fortepiano', 'pianoforte', or just 'piano' as we say today. Instead of plectra, pressing a key sent a small leather-covered hammer upward to strike a string. The more firmly the keys were struck, the more vigorously the hammers did their work and the louder the resulting note. The player now had full command over loud and soft in music-making.

There were other cunning innovations too: an 'escapement', which immediately let the hammer fall back to its starting point, ready to be used again; and 'dampers', small cloth-covered blocks that pressed upon vibrating strings as soon as the keys were released and stopped the sound.

Cristofori may not have been the first to experiment with such a mechanism: records survive of similar designs from elsewhere. But not until Cristofori did anyone follow through consistently with the idea; the oldest surviving instruments (from the 1720s) are his. The pianoforte was not immediately popular: as composers like Bach pointed out, the early instruments were heavy in touch and weak in the high notes.

So the harpsichord remained the keyboard of choice (and used in orchestras) through most of the eighteenth century. But many generations of technical development, starting in Germany and England in the nineteenth century, produced the advanced instruments of today. The most sophisticated 'grand pianos', costing hundred of thousands of dollars, use the principles pioneered by Cristofori. In the Western world, every middle-class household had a piano until recently, and more people would have played the piano than any other instrument.

Dentistry Gets Going

John Greenwood, Pierre Fauchard, Auguste Taveau

Many ancient cultures employed elaborate techniques to replace or repair decaying or missing teeth, but it took a long time to get decent dental care in Europe. The first comprehensive manual on dentistry, Frenchman Pierre Fauchard's *The Surgeon Dentist*, appeared only in 1728. Poor diet compounded the problem, especially once discoveries in the New World brought us refined sugar. Tooth decay became endemic.

Repairing a decayed tooth requires the rotting material to be cut away—never quick, easy or pleasant. Various types of drills were devised late in the eighteenth century, twirled in the fingers or by the 'bow strings' used by jewellers. John Greenwood, dentist to the first US president, George Washington (who had terrible teeth), used a drill powered by a foot pedal. During the nineteenth century, drills were devised driven by air from foot bellows, clockwork or hand cranks. Systems of pulleys or flexible cables carried the rotation from the 'engine' to the drill-piece in the dentist's hand.

Electric motors arrived late in the nineteenth century, but they were heavy and expensive and took decades to become widespread. Foot pedals drove most dental drills well into the twentieth century, leaving the dentist's hands free, but at an often agonising 800 turns a minute. A dentist's patients needed fortitude. Drills spun by air turbines many hundreds of times a second, water-cooled to reduce heating, became commonplace only over recent decades, greatly reducing pain and fear of going to the dentist.

To fill the hole prepared by drilling (by whatever method), all sorts of materials had been tried: stone chips, gutta-percha, tin or lead foil or cylinders, pieces of cork; gold had long been favoured if it could be afforded. Power-driven drills raised the demand for cheap filling material and in 1816 Frenchman Auguste Taveau first mixed silver from coins with mercury to make an 'amalgam' that was soft to work with at first, but hardened when in place. Amalgam is only now being phased out, replaced by synthetic materials.

To reduce the pain of early dentistry, preference was for nitrous oxide (p. 107), inhaled through a mask; it was more effective mixed with oxygen. Cocaine injections could induce local numbness in teeth and gums (p. 110), but its addictiveness drove study and later modification of its chemical structure. By 1905 Novocain (meaning 'new cocaine') was available, powerfully effective but not addictive. Popular later, and still current, is lignocaine.

The Revolution in Textiles

From John Kay to Richard Arkwright

The Industrial Age (or 'Revolution') began in the textile industry. The transformation was surprisingly quick. In only a few decades, traditional home-based spinning and weaving were replaced by factories using machines so fast that textile production increased tenfold in just two decades from 1770. But these were troubled times as well. Many textile workers, fearful for their jobs, tried to destroy the threatening machines. The opposition to William Lee's early 'knitting machine' (p. 22) was even more in evidence.

The Flying Shuttle

In 1733 Lancashire farmer's son John Kay invented the 'flying shuttle' to speed up the weaving of broadcloth. This process normally needed two weavers to throw the shuttle carrying the 'weft' thread back and forth over and under across the bed of 'warp' threads. Kay's machine used a lever pulling strings to do the job, dispensing with one weaver.

Spinning Jenny

The success and speed of Kay's flying shuttle raised the demand for yarn. In 1761 the Royal Society for the Arts offered a prize for whoever could devise a machine able to 'spin six threads of wool, flax, hemp or cotton at a time, while requiring only one person to work and attend it'. The centuries-old spinning wheel could manage only one.

One solution came from the illiterate carpenter and weaver James Hargreaves, with his 'spinning jenny', which reproduced mechanically the hand movements of the user of a spinning wheel. His machine, named after his wife or daughter, or perhaps just a corruption of 'engine', could spin eight threads at once. Later models could handle over 100. The threads were coarse, suitable only for weft, but the hand-spinners were still aroused by the threat to their employment. A few years later they broke into his house and destroyed his machines (they did the same thing to Kay in 1753).

The Spinning Frame

Other hands and minds were at work. In the same year (1764) Richard Arkwright reputedly invented (and patented a few years later) a much larger device he called a 'spinning frame'. This employed rollers to spin and twist, so producing stronger yarn for use as warp. When power from a waterwheel replaced hand operation in this machine, it became known as a 'water frame'.

The Spinning Mule

Samuel Crompton's 'spinning mule' combined features of the spinning jenny and the water frame (hence its name), overcoming some of the deficiencies of both. Other innovations ensured that the spun thread was held under a constant tension. The machine first appeared in 1779, after five years of secret development in the attic of a Lancashire manor house where his widowed mother was the housekeeper. Crompton, who had spun as a youth to supplement the family income, now supported himself playing the violin in a nearby theatre.

Crompton's mule produced thread of excellent fineness and quality, suitable for making muslin for example, the operator having unprecedented control over the process, and everyone wanted to know the secret. But Crompton did not secure a patent. The common practice of the time was to require the inventor of so widely used a machine to make it publicly available in return for a 'subscription', contributed by interested parties. Crompton received merely £60, though in 1812 he was granted £5000 by parliament. By that time, his mules across the counties were producing the output of 4 million spinning wheels, only much faster.

The Power Loom

Clergyman Edmund Cartwright helped the weavers catch up again with the spinners in 1785 with his 'power loom', essentially John Kay's flying shuttle model driven by water or steam power. Unlike Kay, Cartwright ultimately gained both monetary reward and public acknowledgement, including a 'memorial' presented by parliament for his services to the nation, though he went bankrupt at least once.

Apart from Cartwright few of the inventors of these machines made much money from their ingenuity. They struggled to secure and protect their patents and were regularly buffeted by other inventors as well as by protesters. John Kay died in poverty and obscurity in France, and others had to wait years for their rewards.

Factories

One man who did wrest a fortune from the textile revolution was the aggressive and self-sufficient Richard Arkwright, not so much as an inventor (he may have pirated the idea for the water frame) but as an entrepreneur, one of the first. He somehow secured the patents others, such as Crompton, failed to get (though he also lost the water frame patent), invested money in further development, and then put the improved machines to work in the first large-scale cotton mills. Dozens of spinning and weaving machines, powered first by horses, then by water and finally by steam, all worked under the same roof.

In short, Arkwright founded the factory system, made a substantial fortune and changed the world, not entirely for the better. Much cheaper and more plentiful cloth was one legacy; child labour, one of the many horrors of the coming industrial age, was another. Workers servicing the factories would soon be filling the first industrial slums.

The Success
of the Sextant

John Hadley, John Campbell

For eighteenth-century science or commerce, no challenge was greater than finding a way to for a ship's captain to establish his position once out of sight of land. Fortunes hung on the safe arrival of cargoes of raw materials, manufactured goods or slaves, not to mention the lives of the crew. Ships needed to know where they were if they were to reach their destination safely.

The 'global-positioning system' in universal use had been devised 1500 years earlier by the venerable Greek geographer Ptolemy. Any point on Earth lay at the intersection of two lines: a parallel of latitude running east–west aligned with the equator, and a meridian of longitude, running north–south from pole to pole. Identify the lines you are on, and you are where they cross.

Finding latitude was the lesser challenge. The Earth is pretty much a sphere, so travelling north or south changes your view of the sky. It affects how high the Sun, the Moon or a 'fixed star' rises above the horizon to the north or south. For example, the Sun is right overhead at noon on the equator, though only on two days of the year. Navigators have to watch the calendar and use specially prepared tables at other times of the year.

For centuries, sea captains had relied on crude 'cross-staffs' to measure the altitude of the Sun at midday and so figure their distance north or south of the equator. In 1701 the English scientist Isaac Newton suggested that with the right equipment, the same idea could be used for any star whose position was precisely known. The resulting need to accurately map the positions of stars was a major force in astronomy at the time.

But how to do it accurately? The old methods required a captain (often on a heaving deck) to keep one eye on the Sun or star and the other on the horizon, and to measure the angle in between. It was a tough ask. In 1736 John Hadley of England simplified matters greatly with an instrument with internal mirrors. Move those to bring the horizon and the celestial object into the same line of sight, and the angle between them could be read off a scale. His versions could measure angles up to 90 degrees (a 'quadrant') or 45 degrees (an 'octant'). A little later, John Campbell settled on 60 degrees, and so we know the instrument as a 'sextant'.

Finding longitude proved a much tougher problem (p. 50). Now of course satellite systems let anybody find out just where they are (p. 277). 'Yachties' might still keep a sextant, but only for emergencies, for nostalgia or to mystify their passengers.

How Hot, How Cold?

Isaac Newton, Daniel Farenheit, Anders Celsius

Three hundred years ago, the thermometer as we know it now did not exist, though there had been some attempts to measure very roughly how hot or cold things were, such as the primitive air-filled thermoscope invented by Italian physicist Galileo in 1593. He and others knew that air or a liquid such as water expanded (took up more space) as it warmed, and that increase could be used as a measure of temperature.

Leading the way now was Englishman Isaac Newton, the great man in science at the time. In 1701, he greatly improved existing models by marking 'fixed points' on a scale alongside a sealed and evacuated glass tube contain a thread of liquid. He used oil, but water, alcohol and mercury were all possibilities.

Newton chose the temperature at which water froze as the lower one of his fixed points, and the temperature of the human body or blood, which seemed to be roughly constant, as the upper. The length between the two marks was divided equally into 'degrees'. Now the thermometer was accurate enough to be useful.

In 1714 Prussian-born Daniel Fahrenheit made his contribution. His name was to be linked with the measurement of temperature almost to the present day. In his thermometers, he used both alcohol and mercury but settled on the latter because it could measure a much wider range of temperatures. For his fixed points he used the melting of ice at (32 degrees on his scale) and the boiling of water (212 degrees), dividing the interval between them into 180 equal steps. It is not clear why he chose such inelegant numbers but the zero point on his scale was reputedly set by the lowest temperatures he experienced in Amsterdam, where he had settled.

In recent times his temperature scale has been almost universally replaced by the Celsius or 'centigrade' scale but it lingers in use in the United States and among older people in other countries. Anders Celsius of Sweden devised his scale in 1742, with water boiling at 0 degrees and ice melting at 100 degrees. This was soon inverted to provide the temperature scale now in almost universal use. Celsius's increasingly accurate mercury-in-glass thermometers became widely used in laboratories. By the 1770s they could read temperature to one-tenth of a degree. Such precision became increasingly important in both science and invention.

Among the vital discoveries made possible were 'latent heat' and 'specific heat'. These greatly increased the efficiency of steam engines (p. 51).

The Scourge of Scurvy

James Lind

Scurvy was rightly feared by eighteenth-century mariners, captains and crews alike. It caused physical weakness and depression, livid spots on the skin and swollen and blackened gums. Untreated, the worst cases died, yet once ashore fresh food could clear the symptoms in a matter of weeks. Some treatment or cure was needed; able to be applied at sea, able to prevent the heavy burden scurvy was laying on crews and passengers of both naval and commercial ships.

In 1747 English naval doctor James Lind was aboard HMS *Salisbury*, patrolling the English Channel during the War of the Austrian Succession. In spite of good food and 'sweet water', scurvy broke out. Within 10 weeks at sea, 80 men out of 350 were down with the disease. Lind knew the juice of citrus fruits such as oranges, lemons and limes had long been suspected of being an effective remedy. So he designed an experiment to see how the 'citrus treatment' compared with other remedies.

Lind chose 12 sailors already suffering from scurvy, and grouped them in pairs. Each pair had a different addition to their normal rations of gruel, mutton broth and ship's biscuit. These included spoonfuls of vinegar, cups of seawater, a mash of garlic, mustard and horseradish, and a quart of cider. One pair had an 'elixir' of complex composition; the last pair had had two oranges and one lemon each day. Only cider and the fruit caused any noticeable improvement, and the fruit far the most. One of the men was fit again for duty in six days.

The findings were very clear, and quickly accepted by the Admiralty. Juice of lemons and oranges (and limes) was the best cure for scurvy. Yet 40 years passed before an official order was made to supply lemon and lime juice to all Royal Navy ships. Within another year, scurvy had all but disappeared and the number of sailors hospitalised for any cause was halved. Another consequence was the adoption of the nickname 'limey' for English sailors, and ultimately for all English abroad.

Most fundamentally, a specific disease had been linked to a factor in the diet of the sufferers. The identification of a 'vital amine' as the missing factor, what we now call vitamin C or ascorbic acid, would take another two centuries but the way was now open for similar action to be taken for other conditions.

A Man of Many Parts

Benjamin Franklin

In 84 years of life, American Benjamin Franklin tried just about everything. He was publishing the *Pennsylvania Gazette* at the age of 23, and *Poor Richard's Almanack* at 32. He founded a circulating library and several fire insurance companies, organised a militia, and became Postmaster of Philadelphia.

In his mid-thirties, he turned to science, especially experiments with electricity. Politics steadily took up more of his time, first local, then national. At 70 a signatory to the Declaration of Independence, he was later ambassador to France and was in Paris to see Jacques Charles go aloft in the first hydrogen balloon (p. 57). He helped negotiate peace with Britain in 1782, and campaigned against slavery.

His active mind teemed with invention. Some came from his own needs. Both longsighted and shortsighted, he needed different spectacles to read and for distance vision. Frustrated by the constant changing, he cut the lenses in two horizontally, attaching the bottom part of his reading glasses to the top half of the spectacles for everyday. Franklin had invented 'bifocals'.

As postmaster, he needed to know the length of the various postal routes to aid his planning; an 'odometer' he invented for this purpose was attached to his carriage. His brother John suffered from kidney stones, so Franklin developed a flexible urinary catheter, one of the first ever.

Noting that the everyday fireplace sent much heat into its surrounding wall rather than into the room, Franklin built the cylindrical cast-iron 'Franklin Stove' to sit in the centre of a room, sending out heat in all directions. His version had no chimney; the fire would not 'draw' and often went out. Fellow Philadelphian David Rittenhouse added an L-shaped flue to vent the smoke into a nearby wall-chimney, making a draft to keep the fire alight.

Franklin's famous 'kite experiment', supposedly in 1752, proved that lightning is a huge electrical spark. This prompted his invention of the 'lightning rod'. He claimed a pointed metal rod would draw electric charge from nearby clouds before it built up enough to generate a lightning strike.

The first lightning rod (or 'conductor'), set above the roof and connected by a wire to the ground, was on his own house; soon after, they appeared on major public buildings. The invention won Franklin a medal from the Royal Society of London, and soon after he became a Fellow. That made him a top man in science to go with his other achievements.

Improving the View, Inwards and Outwards

John Dolland

Scientists using microscopes and telescopes in the mid 1700s were frustrated by the imperfections of even the best instruments of the time. One problem was the blurred coloured fringes that surrounded the image, making good observations difficult. This 'chromatic aberration' was the result of the different colours of light being bent by different amounts as they passed through the lenses. It was to defeat this problem that Isaac Newton pioneered the reflecting telescope, which uses a mirror rather than a lens to gather light (p. 23).

The hero of the hour in this matter was John Dolland, son of a Huguenot weaver in east London. He was a weaver for a time himself until he set up in business in 1752 with his son as a maker of optical instruments. The answer to the colour fringe problem had been suggested: use combinations of materials like glass and water in the lenses, as these bent light by different amounts and, if properly arranged, would cancel out the effect.

The idea seems to have come from one Chester Moor Hall, a 'gentleman' of Essex. He argued that the different 'humours' of the human eye so refract rays of light as to produce an image on the retina that is free from colour. Perhaps combining lenses composed of different materials might do the same. As a man of independent means, Hall was perhaps indifferent to fame; at least he took no trouble to communicate his invention to the world.

When he heard Hall's suggestion, Dolland had some doubts. It seem to conflict with Newton's influential ideas, but he began to experiment. He soon found success with the water/glass combination, but combinations of different glasses, such as 'crown glass' (window glass) and 'flint glass' (p. 32) were more practical. By 1758 his 'achromatic' lenses was free of the troublesome coloured fringes. He gained the gratitude of observers, and also the Copley Medal from the Royal Society. He was a Fellow two years later and in 1761 he became optician to the king.

Science benefited too. His invention brought telescopes with lenses back into favour, as these made better use of the available light. Biologists were comfortable again with multi-lens ('compound') microscopes; within half a century these were able to magnify a thousand times. Both the macrocosm and the microcosm began to yield more secrets.

The Reinvention of Cement

John Smeaton, Joseph Aspdin

Eddystone Lighthouse warns ships of treacherous rocks off the Cornwall coast. The first lighthouses there, built in 1698 and 1709, were wooden structures, soon destroyed by storm or fire. Yet the third lighthouse, finished in 1759, operated for nearly 130 years. When erosion of the underlying rocks required its demolition, the lower levels were so firmly bonded they could not be taken apart. They remain in situ to this day.

One secret to that strength was 'hydraulic cement'; that is, cement that sets under water, even salt water. Its inventor, or rather its reinventor, was John Smeaton. The ancient Romans had known that 'quick lime' (made by heating limestone) mixed with a larger amount of a certain crushed volcanic ash made a cement that did not need to be dry to harden, unlike ordinary mortar. This knowledge was lost at the fall of Rome and not rediscovered for 1500 years.

Smeaton, already eminent in 'civil' (as opposed to 'military') engineering, had been recommended by the Royal Society to build a new lighthouse at Eddystone. To storm-proof the 20-metre (66-foot) tower, he employed three new ideas: a conical shape (modelled on an oak tree), interlocking blocks of stone, and his rediscovered hydraulic cement. Drawing on records of Roman experience, he had experimented with various mixes of lime and clay (such as from crushed bricks) in search of the ideal formula. The durability of the lighthouse demonstrates that he found it.

Smeaton's innovation was taken forward by others in search of a consistent product. In 1824 Leeds bricklayer and stonemason Joseph Aspdin patented a form of hydraulic cement called 'Portland cement'; named because it resembled the colour of stone quarried at Portland on the English South Coast (and when set was as hard).

Aspdin crushed and heated together precise amounts of limestone and clay, grinding the result into cement. A similar process is used today, when concrete (cement mixed with sand or gravel) is the world's most widely used building material. Early successes with Portland cement included repairing the first tunnel under the Thames in 1828 and cementing the millions of bricks that formed the new London sewers in the 1860s.

Smeaton's achievements beyond hydraulic cement were significant and diverse: discoveries in physics, improved water wheels and windmills, many bridges, canals and harbours. Like other industry leaders, including James Watt, Matthew Boulton (p. 51) and William Murdoch (p. 69), he was linked with the famed Lunar Society of Birmingham.

Discovering the Longitude

John Harrison

In 1707 four Royal Navy ships and 2000 men were lost in a single shipwreck, mostly because they did not know where they were. A ship's captain could find his latitude quite easily (p. 44) but reliably finding longitude east or west of his home port had so far proved impossible.

Taunted by such losses, the British parliament passed the *Longitude Act* in 1714, offering to reward 'such person or persons as shall discover the Longitude': £20 000 if the longitude of a ship at sea could be established within one degree, £40 000 if within half a degree. These were immense sums for the day, but the problem was urgent.

The solution had long been known, at least in principle. The captain simply compared his local time set by the position of the Sun in the sky with the time (at that same moment) back in the port he had come from. Since the Earth turned through 360 degrees every 24 hours, each hour difference in time meant 15 degrees difference in longitude. But just what time was it back home?

There were two possible ways to find out, both long understood, both with problems. You could take a clock with you from the place you had left, but would it be reliable enough? Champion clockmaker Christiaan Huygens had tried that route and failed (p. 29).

Or you could use a 'clock' that would be visible from both places, say, something in the sky. For example, the movement of the Moon in front of stars, though complex and not quite regular, could be likened to the hour hand against the numbers on a clock. The latter 'method of lunar distances', which used John Campbell's sextant (p. 44), proved hard to compute and to use. It required many pages of mathematical tables devised by the Astronomer Royal, Nevil Maskelyne, who championed the method and published in the *Nautical Almanac*. It did, however, let the above-average captain find his longitude to within a degree (about 100 kilometres).

Yorkshire carpenter-turned-clockmaker John Harrison did much better. He took the first route, and spent 40 years on the task. By 1765 he had done the impossible: he built a clock so reliable it lost or gained less than a minute in a six-week transatlantic voyage. This gave the longitude well within the acceptable error. Knowing the time to the nearest minute meant longitude accurate to a quarter of a degree. Envious manoeuvrings among the members of the 'Board of Longitude' scandalously denied him his reward for years, but with the intervention of the king, 'Longitude Harrison' finally got the money and his place in history.

Getting More from Steam

James Watt, Matthew Boulton

Many will say that James Watt invented the steam engine. But steam-enabled 'atmospheric engines' (p. 37) had been pumping water from mines for many years before the Scots-born Watt first took an interest. These worked but were horribly inefficient, wasting 99 per cent of the energy of the coal burnt to raise the steam.

Asked to repair a model Newcomen atmospheric engine in 1769, Watt soon saw a major reason for the waste. The cylinder and piston had to be cooled during each stroke to condense the steam, and heating them up again absorbed some of the heat in the next steam blast. Watt's innovation was to build a separate condenser, so the cylinder could always be kept hot. Science had played a role here: studies by Watt's countryman Joseph Black had measured the energy needed to heat the cylinder each time, energy that essentially went to waste.

Watt had some other very useful ideas: the 'governor' to keep the engine running at a steady pace; and 'sun and planet' gearwheels that let the engine drive rotating machinery, like the new large-scale spinning and weaving equipment coming into use. Watt also had a business partner, Birmingham iron-maker Matthew Boulton. By the 1780s, the pair enjoyed a good living manufacturing the engines that drove the early Industrial Revolution.

Watt's engines still could not pull their own weight continuously, so there was no chance of a reliable self-moving steam engine, a 'locomotive'. Progress demanded the use of higher pressure steam to drive the piston, without any help from air pressure. This meant higher temperature working as well, extracting more energy from the steam. Efficiency climbed steeply; steam engines became lighter and more compact.

There were hazards too, early boilers could not withstand the increased steam pressure and often burst, with catastrophic results. One precaution was the 'safety valve', invented a century before by Frenchman Denis Papin for his 'steam digester' (aka pressure cooker); this released steam if the pressure rose too high (p. 34). Acceptable safety came only with much experience, better design, materials and manufacture and, ultimately, government regulation.

Names to note include Englishman Richard Trevithick (p. 83), whose high-pressure 'Cornish engines' were successful and popular in factories and mines. An additional refinement introduced 'double-acting' engines, with steam pushing alternately on both sides of the piston. It was now time for the first steam-powered vehicles (p. 52) and boats (p. 81).

Steam on Wheels

Nicolas-Joseph Cugnot

Nicolas-Joseph Cugnot's 'steam wagon' (*fardier à vapeur*) was probably the first 'horseless carriage'. This French army officer was first to put a steam engine on wheels so it could pull loads, such as heavy artillery pieces, as horses had done for millennia. By 1770 he had a version, a three-wheeler, steered by tiller like a boat, able to pull 4 tonnes at 4 kilometres an hour.

Steam engines of the time lacked power and efficiency, with important advances by James Watt and others still to come. So Cugnot's steam cart regularly ran out of steam and had to stop every 20 minutes to let the pressure build up again. The vehicle was also the subject of the first recorded motor vehicle accident, running off the road into a garden wall. A second improved model was never seriously tested, though it has been on show since 1800 in a Paris museum.

Cugnot, too, had troubles. He lost his pension when the Revolution broke out in 1789, and lived in exile and poverty in Brussels. Napoleon pensioned him again, just before Cugnot died in 1804, but showed little interest in the machine—surprising given the Emperor's well-known technological awareness.

No one in France followed up Cugnot's work. Across the Channel, pioneers like William Murdoch (p. 69) and Richard Trevithick (p. 83) were working with steam locomotives, but interest was mostly in railways. Steam engines were still too heavy for the poorly made roads.

Around 1807 Swiss Isaac Rooz tried another tack, successful in the long run. He replaced the steam engine, with its boiler and fire, with an 'internal combustion engine', perhaps the first, other than the failed experiments by Huygens and Papin with engines powered by exploding gunpowder. He drove the pistons in his engine not with introduced steam but with hot expanding gas made by igniting a mixture of hydrogen and oxygen (this was actually steam too, of course). Around 1813, his last model, 6 metres long and weighing a tonne, could barely move itself.

Fifty years later, French inventor Etienne Lenoir tried again, using coal gas as fuel. The engines themselves worked well and were popular with small factories not needing a large steam engine. From around 1860, he sold 500 in the Paris area alone. Again they were too heavy and inefficient for a practical vehicle but by then the first successful 'automobiles' were not far away (p. 146).

Chemicals Old and New

James Tennant, John Roebuck

Chemistry came of age in the late eighteenth century. Brilliant discoveries in England, Scotland and Sweden culminated in the new chemical theories of the French genius Antoine Lavoisier. We began to understand how and why chemicals react as they do. At the same time, chemistry became more useful. Ways were found to make large amounts of chemicals of value in industries like the manufacture of textiles, glass and paper, often replacing substances derived from nature that had been used for centuries. Among the most vital of these were chlorine and sulfuric acid.

Chlorine, a green gas released from hydrochloric acid, had been discovered by the industrious Swedish chemist Karl Scheele in 1774, just before the American Revolution. We soon found that chlorine bleached the colours from cloth or other natural materials But chlorine is nasty stuff to handle, poisonous and corrosive, and therefore not safe for routine fabric whitening and stain removal. Its use was limited.

It was Scotsman James Tennant who found a way to render chlorine safe to handle, and therefore genuinely useful. He passed chlorine over lime (calcium hydroxide), making solid 'bleaching powder', which could be stored until needed. In water, bleaching powder releases chlorine. Suddenly, dazzling white cottons and linens were easily available; the textile industry, already mechanising at a fast rate (p. 42), got a tremendous boost.

The traditional way to bleach cloth was slow and laborious, involving repeated steeping of cloth in sour milk (a mild acid) and exposure to sunlight for weeks at a time. The Dutch were good at this and had dominated business. Now anyone could do it, thanks to bleaching powder. The same process whitened paper, removing natural colours from the vegetable fibres.

Sulfuric acid was another chemical coming into vogue. When diluted, it was a satisfactory bleach; in concentrated form it could refine gold and silver. Sulfuric acid had been made in small amounts for hundreds of years, for example by burning sulfur and mixing the resulting gases in water in glass vessels.

Around 1760 English chemist John Roebuck began to run this process in large chambers made from lead (immune to the corrosive acid) and produced the acid much more cheaply and in greater quantity than ever before. Sulfuric acid became the most important industrial chemical, used in almost every chemical manufacturing process. Until the late twentieth century, a nation's level of industrialisation could be measured by its production of the stuff.

Masters of Iron

Abraham Darby, John Wilkinson

In 1781 the first bridge made entirely of iron opened for traffic across the Severn River in western England, close to Coalbrookdale. The arched bridge, 15 metres long and 14 metres high, had taken three years to assemble. Nothing like it had ever been built before.

Nearby, in the midst of a rural landscape, stood the state-of-the art iron works run by Abraham Darby, a third-generation ironmaster (with a father and grandfather also called Abraham). Here, all the iron in the bridge had been cast and shaped. Darby's partner in this enterprise was an ironmaster of equal repute, John Wilkinson, whose name lives on in Wilkinson's Sword razor blades. Together they symbolised iron, the dominant material of the looming industrial age.

In 1709 Darby's grandfather had pioneered the use of coke, obtained from coal, in place of wood-derived charcoal for the release of molten iron from its ores in furnaces powered by bellows. Suitable wood had become expensive, with the wholesale felling of trees to meet the demand for iron weapons, necessitated by the regular convulsion of Europe by war. By the end of the century, ironmasters would be making quality iron in large amounts using coke, and more of it was for peaceful pursuits—building, transport and manufacturing.

The iron in the bridge and other applications had begun as 'cast iron', the raw output of the furnaces, run into moulds of sand to cool and harden. It was strong enough for many uses, but brittle. Reheating and hammering could strengthen it but no one had dared to make a substantial structure from the stuff, for fear of catastrophic collapse. A few decades later, in the form of 'wrought iron', it was everywhere.

This second Iron Age was also the Age of Steam. Wilkinson played a key role there with his 'boring machine' that cut the cylinders of steam engines from a block of iron. Making these adequately round and smooth internally was beyond the established techniques, such as casting and then chiselling and filing. Wilkinson's machine was robust, holding the iron block firmly in a cradle of oak beams. Water power drove the boring bar around and around. So well did the machine perform that Wilkinson's friends, the steam engine makers James Watt and Matthew Boulton (p. 51), relied on his borer for 25 years.

'The Author of the Iron Aristocracy'

Henry Cort

As an example of an industry pioneer who enriched others and the nation as a whole, yet received almost nothing in return, it is hard to go past Henry Cort. Son of a Lancashire brickmaker, he set up business aged 25 in London as a buying agent for the British Navy, sourcing, among other things, increasing amounts of iron for weapons, anchors and chains. Cort became aware of how poor British iron was compared with imports from nations like Russia and Sweden. (Knowing how much Britain needed its iron, Russia had no hesitation in raising the price two and threefold!).

After 10 years working with the navy, and having accumulated a little capital, Cort left London to set up his own iron-making business on the shores of Portsmouth Harbour, close to the naval base. Here he began to experiment with new techniques to make stronger, tougher iron, able to compete with imports.

He first developed a process to roll out iron bars from the crude 'pig iron' which came from the furnace, rather than hammer them, which was much slower and more labour intensive. The process, using grooved rollers powered by the new steam engine, squeezed out many impurities and spread traces of carbon left from the smelting process evenly through the iron, increasing its strength and reducing its brittleness.

Cort's second innovation converted the crude 'cast iron' into 'wrought iron', which was much more useful and durable, by a process called 'puddling'. A vat of molten iron, heated from above, was constantly stirred by a worker with a long iron bar. Many of the carbon impurities were burned off and the pure iron was formed into a ball on the end of the bar, ready to be hammered into a 'shingle' and then rolled.

These innovations were not Cort's alone: he built on the ideas of others, including the Grange brothers, John Roebuck and Peter Onions, but his own contributions were vital and the outcomes of immense importance. In 1786 Lord Sheffield proclaimed that the new processes meant much more to the nation than the possession of the American colonies (then in the process of being lost), as they would give Britain command of the iron trade 'with its vast advantages to navigation'. The following year, tests on iron produced by Cort's methods found it superior to imported iron, and imports promptly ceased.

Some figures tell the story. By 1820, 8000 of Cort's wrought-iron furnaces would be running in Britain, producing 400 000 tonnes annually, as opposed to a mere 90 000 in 1780. By 1900, production would total 4 million tonnes, more than all other European nations combined. The nineteenth-century preeminence of British industry ('the workshop of the world') was secured, and many ironmasters made fortunes.

Typical of these (and one of the early movers) was Richard Crawshay, another enterprising Yorkshireman, who had ridden to London at the age of 15 to seek his fortune (the journey took a fortnight over the appalling roads of the time). He found employment in an ironmonger's shop, and specialised in the selling of 'flat irons' to washerwomen. He inherited the business when the owner died and soon expanded his interests into iron-making, taking a lease over deposits of iron ore and coal and a small ironworks in South Wales. Profits from the store in London were ploughed into the new venture.

In 1787 Crawshay struggled to make 10 tonnes of iron bars a week, and of poor quality at that. Needing to build new furnaces, he decided to adopt Cort's methods, with such success that a quarter of a century later he could report making 200 tonnes a week, and had built a canal to get his produce to port in Cardiff.

Cort himself did not prosper. Early on, he had entered into a partnership with Samuel Jellicoe, whose father Adam was Deputy Paymaster for the Navy. The elder Jellicoe financed the iron-making experiments, holding Cort's patents on his two processes as guarantee for the debt.

Adam Jellicoe was in debt himself when he died in 1789. To settle the estate, Cort's debt was called in, but he could not pay. His patents, now of vital importance to the navy, were taken over by the government, in part because some of the funds Jellicoe advanced to Cort were moneys entrusted to him to pay the wages of officers and seamen. As no royalties had been paid by the many ironmasters who had made use of his inventions (partly because they disputed the patents), Cort was ruined.

Forced to leave his business in the hands of the younger Jellicoe, Cort was ultimately reduced to seeking a government pension to support his wife and 12 children. Later to be described as 'the author of the iron aristocracy', he died in 1800, broken in health and in fortune.

Up, Up and Away

The Montgolfier Brothers, Jacques Charles

It's not known who first had the idea that humans might fly aboard a hot-air balloon, such as the Chinese had made for millennia, but it may have been Frenchman Joseph Montgolfier. One story has him visualising soldier-carrying balloons storming the British-occupied Rock of Gibraltar. Montgolfier and his brother Jacques began experimenting in 1783, with balloons of linen and paper carrying a fire beneath.

Their first flight involved a hot-air balloon alone; the next carried some passengers, a sheep, a duck and a rooster, according to the popular story. The duck broke its neck on landing after the 15-minute flight, but the other animals survived, proving that the air up there was fit to breathe. In November that year, in front of the king, Francois Rozière and Francois Laurent, brave men both, lifted off in a Montgolfier balloon. They reached 1000 metres and travelled 8 kilometres in 20 minutes before landing safely. Human flight was possible.

The French Academy of Science asked rising star Jacques Charles to investigate. Charles reported that filling balloons with the light gas hydrogen would be safer than relying on a fickle fire. In December he went aloft himself in a balloon covered in varnished silk, the hydrogen provided by sulphuric acid (p. 53) reacting with iron. The balloon rose from where the Eiffel Tower now stands in Paris. Among the spectators was Benjamin Franklin (p. 47), then the American Ambassador to France.

Two years later, Frenchman Jean-Pierre Blanchard, a pioneer of the parachute (p. 73), and American John Jeffries crossed the English Channel in a hot-air balloon, the Channel being the first obvious challenge. They barely made it. In the same year, Francois Rozière died attempting the same feat. He had (unwisely) thought a hydrogen balloon and a hot-air balloon in tandem would provide better lift than either alone. The fire touched the close-by hydrogen balloon, setting the gas alight; disaster followed.

Hot-air balloons remain popular today for recreation and for feats such as crossing an ocean or orbiting the planet. For safety reasons few balloons use hydrogen nowadays, other than the balloons sent regularly aloft by weather observers. Larger gas-filled balloons nowadays use helium, not as light, not renewable, but also not flammable. Such balloons have carried instruments (and humans in sealed capsules) to great heights, almost to the edge of space.

Roads Fit for Travel

John McAdam, Thomas Telford

Road-making is an ancient skill. The Roman conquerors built an excellent network of roads throughout Europe, but 2000 years later these had fallen into disuse and decay, and the construction techniques were forgotten. Roads in eighteenth-century Britain were noticeably bad, so appropriately the first modern road-builders became active there.

Scotsman John McAdam is most often mentioned: roads built according to his methods were said to be 'macadamised'. McAdam had made a small fortune in his twenties and thirties in his uncle's New York counting house. He purchased a substantial estate in Ayrshire on returning in 1783. Problems encountered in moving around his estate led him to consider how roads, mostly little more than rough dirt tracks, might be improved.

But others went before him and beside him, such as the blind Englishman John Metcalfe and Scot Thomas Telford, McAdam's contemporary, a brilliant, versatile engineer. Clearly a solid foundation was needed, stone blocks with the gaps filled with small durable stones or chips of granite rammed into place. For Metcalfe, that was enough; he built hundreds of kilometres of such roads in Lancashire and Cheshire in the late eighteenth century.

Telford and McAdam went further, completing the road with extra layers of stones of decreasing size, ending up with tiny fragments or gravel on top. All this was pressed into place with heavy rollers to make a smooth surface, and water sprayed on to increase bonding. Such roads needed a lot of manual labour, but were durable. The curve or 'camber' on the surface ensured that rain ran off into side drains and did not soak in. Roads were better built once the steam-powered road roller or 'steamroller' was invented by Thomas Avery, formerly a farmer from Kent, 30 years after McAdam died in 1836.

By that time, Telford, McAdam and many others using their methods had built thousands of kilometres of 'highways' throughout England and Scotland, speeding travel by coach (still very slow by modern standards). Similar efforts were underway throughout Europe, the Americas and elsewhere.

Such 'unsealed' roads became dusty or soft depending on the weather. Spraying the surface with a liquid derived initially from coal tar made it waterproof, binding in the dust, and producing a 'tar-macadam' road or 'tarmac', a term commonly used today for aircraft landing strips even when these are made of concrete as many modern roads are.

Safety Locks and the Hydraulic Press

Joseph Bramah

It was unfortunate for 16-year-old Yorkshire farm-boy Joseph Bramah that he suffered an ankle injury that left him unfit for farm work. But it was good fortune for the rest of us, as it set him on to a new career of invention, the fruits of which we still enjoy.

Bramah turned to carpentry, became apprenticed to the village cabinetmaker, and then, despite his lameness, walked to London to seek employment. Soon after, he had his own small business, fitting 'water closets' (p. 24). Bramah quickly improved these with a new sort of valve to seal away the wastes for discharge. The 'Bramah WC' is still found in small boats.

The next challenge was a lock that could not be opened except with a key. His model, patented in 1784, reputedly had 500 million possible combinations. To publicise his lock, he set one in the window of the shop, offering £200 to anyone who could 'pick' it. No one did for 67 years; then it took a skilled American locksmith 50 hours.

During the lock exercise, Bramah first employed the young and dexterous Henry Maudslay, making him foreman soon after. Maudslay was responsible for much of Bramah's success: the locks would work only if made with great precision, as Maudslay had the skill to do. Nonetheless the relationship was to end badly (p. 77).

The Hydraulic Press

The Bramah invention still much in evidence today is the 'hydraulic press', which he created in 1785. He was first to see an application for Pascal's Principle, expounded a century earlier, which states that pressure applied to a liquid is transmitted all through it undiminished. If you multiply the pressure by the area it is applied over, the result is a force. If a small force is applied to the small end of a funnel-shaped tube, a greatly magnified force is available at the large end. That is the basic principle behind the hydraulic press or lift. Very large loads can be lifted and great forces exerted, up to thousands of tonnes, with little effort. The engineer Robert Stephenson would use one to raise the 1000-tonne spans for his Britannia Bridge. It was a great idea, but again needed Henry Maudslay's skills to make it work.

There were many other good ideas; at one time Bramah held 20 patents. He supplied the Bank of England with a machine to print sequentially numbered banknotes. He devised a beer pump, a fire engine and a quill cutter for making pens. He was an early advocate of screw propulsion for ships (p. 99).

Wallpaper and Lino

Christophe Oberkampf, Frederick Walton

Though the Chinese were pasting rice paper on walls for decoration 2000 years ago, wallpaper in Europe was initially a cheaper substitute for cloth hangings, replacing in turn earlier costly tapestries. All wallpaper was hand-painted at first, some resembling marble or stucco. Movable panels allowed owners to take their wallpaper from house to house, like furniture.

Fragments of European wallpaper survive from the early sixteenth century. From 1675 French engraver and major wallpaper innovator Jean Papillion used carved wooden blocks to print matching, continuous patterns. 'Flocked wallpaper' resembling like velvet became popular. A century after Papillion, wallpaper could resemble scenery, needing up to 5000 wooden blocks for printing.

Mechanical printing cut costs and increased popularity. The French led early. Christophe Oberkampf built a wallpaper press around 1785; his countryman Nicholas Robert produced continuous paper rolls for it soon after (p. 75). In Britain, a four-colour wallpaper printing machine, adapted from one printing calico and able to make 400 rolls a day, was patented in 1839. By 1874 wallpaper could be printed in 20 colours. Paper hanging was simplified a decade later with the development of the first ready-to-use paste.

Wallpaper blends artistry with technology. In recent decades its usage has been rejuvenated by the innovation of washable, stain-resistant vinyl materials (p. 214), prepasted papers and the growth of digital technology that reproduces almost any image, including classic designs.

Covering the floor was more expensive than hiding the walls. Carpets were restricted to the houses of the wealthy; floorboards were often bare, cold and dusty. In 1860 Scotsman Frederick Walton pioneered a floor covering called 'linoleum', the name reflecting the use of linseed oil. The capacity of linseed oil to form into thin sheets as the molecules linked up during drying was already being exploited in paint.

Walton started with a woven backing of cotton or flax, adding many layers of linseed oil thickened by boiling and mixed with cork dust. It was repetitive and slow and therefore expensive, and required large sheds for drying. Not till around 1900 did Walton discover how to speed up the process, including artificial heating to dry and harden the material. Cost went down and usage went up, especially once colourful designs could be printed.

The name 'lino' is still familiar but the product is mostly obsolete, supplanted by the ubiquitous 'vinyl' from the twentieth-century boom in synthetics.

A Stick and a Ball

Cricket, Baseball, Tennis and Other Games

Games in which a ball is struck with a stick are popular worldwide. The 'stick' can be called a bat, a club, a racquet or a paddle. The objective may be to score a 'goal', to win a 'point', to hit a 'run'. Matches can be between individuals or teams. The ancestors of these games may have been associated with ancient religious rites marking say, the start of spring. Yet most of them, like football (p. 119), were 'invented' (or at least took their modern forms) only in the eighteenth or nineteenth century.

Cricket

The origins of cricket lie many centuries back, and games between teams of 11 were played for high stakes in the seventeenth century. However, the key date is 1787, with the founding of the Marylebone Cricket Club (MCC) in London. The MCC drafted a set of rules, including the 22-yard pitch and specific ways for players to be dismissed, and remains dominant in rulemaking today. In the nineteenth century, games involving county teams proliferated; Australia beat England in the first international 'Test Match' in 1877. The game spread through the colonies of the British Empire and remains strongest in those nations today.

Baseball

Legend claims that Abner Doubleday invented baseball in Cooperstown, New York, in 1839, but in fact the 'the great American pastime' evolved from existing English games such as rounders. In 1846 the so-called 'Knickerbocker Rules' established the basic elements: nine-player teams and the four-base diamond. Over the next few decades, contests became limited to nine innings, national leagues were founded and professional teams took the field. The latter established that sports could be as much for entertainment as for participation, for spectators as much as for players. It was a new idea.

Hockey

Of the 'score-a-goal' games, hockey has ancient origins in Egypt, Persia and Greece, but developed its modern form in Europe, mostly in England. The Blackheath Club in London, the first ever, was established mid-nineteenth century. Like cricket, hockey spread throughout the British colonies; nations on the subcontinent have a notably strong reputation. In the 1870s W. F. Robertson, a student at McGill University in Canada, set down the emerging rules of a form of hockey played on frozen ponds with players on skates. Today, some see ice hockey as a game with few rules.

Lacrosse and Hurling

From Canada also comes lacrosse, a Canadian Indian game renamed from the supposed likeness of the netted stick to a bishop's crossed staff *(la croix)*. The rules were first written down in 1863. 'Hurling' is barely known outside Ireland, but adherents declare its origins ancient. Rules were first agreed in the 1880s and the annual All-Ireland Championship held since that time.

Golf

Golf is perhaps unique in needing no opponent, though there usually is one. The contest can be just with the course or with your previous best score. With antecedents dating back to the 'paganica' played by Roman emperors, 'golfe' emerged in Scotland in the fifteenth century. The Royal and Ancient Golf Club at St Andrews, established in the 1750s, maintains a dominant influence, but the global expansion of the game began only in the late nineteenth century, when the USA established its first golf courses and national championships.

Tennis and Other Racket Sports

Something akin to tennis was popular among seventeenth-century French aristocrats, though it was often played with the palm of the hand rather than a racket. The name comes from the French for 'Receive!', as the serving player was required to shout. 'Royal' tennis was already popular in England, played indoors with some rules like squash. Modern 'lawn' tennis was invented in 1873 (and patented a year later), by Major Walter Wingfield as a modification of royal tennis. It was first played at a garden party at his home in Wales, initially on an hourglass-shaped court with a very high net. Quickly established among the well-to-do in the USA, its popularity boomed and has never waned.

Squash ('squash rackets') was devised at Harrow School, England, before 1850. Table tennis emerged in the late nineteenth century but its origins are obscure. Badminton or 'shuttlecock' was brought home from India by British Army officers, including the Duke of Beaufort. He played it at his country estate at Badminton, hence its name. The Badminton Association of England, established in 1893, still controls the rules.

Sport and Technology

Evolving technology has benefited all these sports. Sticks and balls may contain advanced materials such as carbon fibres; scientific design now enhances performance. Information technology often helps control the progress of matches (for example, line-calls in tennis) and speeds the scoring; the increasing reach of mass communications can bring a game 'live' to billions of listeners and viewers across the globe.

Threshing the Grain

Andrew Meikle

Andrew Meikle's gravestone in Prestonkirk, not far from Edinburgh, where he died at the age of 92, tells us he was descended from 'a long line of ingenious mechanics'. Certainly his father had invented a 'winnowing machine' to separate the grains of, say, wheat, from their stems, though the common hostility to machines at this time (around 1710) meant that it was not well received.

Andrew did rather better. His 'threshing machine' (or 'thrashing machine') did more than separate the wheat grains from the straw: it actively shook the ears of wheat and liberated the grains, a task until then done by hand. His was not the first such machine, but previous attempts had had limited success, apparently (or so Meikle thought) because they tried to rub the grain from the ear, as a human hand would do. His machine beat the ears of wheat against a drum to set the grains free, much as the fibres of flax (the starting point for linen) were broken from the plant stems.

Meikle built his first model around 1778, patented it in 1788 and started manufacture soon after. At that time (and indeed for most of his life) he was employed as a millwright at Heston Mill, where wind power drove the grinding of wheat into flour. The owner of the mill was John Rennie, whose family estate 'Phantase' was close by. Rennie supported Meikle, putting the new technology to work in mills other than his own. It is said that Meikle's inspiration contributed to Rennie later becoming a leading civil engineer.

Mechanical threshing had come to stay, adding to the boost in agricultural productivity from the use of machines (p. 94). The early machines were powered by wind, but they could also be driven by water, by the newfangled steam engines or, if all else failed, by horses.

As a millwright, Meikle made other improvements, most noticeably replacing the traditional canvas sails of the windmill with a series of shutters that could be opened by levers. The mill would now be closed down quickly in the event of damaging strong winds.

Meikle's inventions did not make him rich or even comfortable. Late in life, his lack of means caused the former President of the Board of Agriculture, realising what this pioneering 'agricultural engineer' had done, to raise £1500 to support him.

In Telescopes, Size Matters

William Herschel, The Earl of Rosse

In the many decades following Galileo's pioneering observations (p. 23), astronomical telescopes advanced mightily in size, power and sophistication. The troublesome coloured fringes that had frustrated early observers were gone (p. 48). Telescopes with lenses were the first choice of most astronomers, since they wasted less light. Nonetheless, there was a limit to how big a refracting telescope could be. Supported only around its edge, a very large lens would collapse under its own weight.

If size mattered, then the 'reflecting' telescopes that Newton had pioneered, with mirrors supported from behind, clearly had an advantage. And size did matter. Bigger telescopes gathered more light. Double the 'aperture' of a telescope and you could see twice as far, and discern details half the size.

German-born English astronomer William Herschel led this drive for aperture. It is not clear why he switched from his early employment in music to astronomy, but by 1776 he was building his own telescopes. With his first instrument, 2 metres long, he found the planet Uranus in 1781. With financial support from the king, he doubled the size and doubled it again, finishing in 1789 with a monster 12 metres long, bearing a mirror more than a metre

across. Constructed in Slough, west of London, the scaffold and turntable used to support the telescope and point around the sky were bigger than a house. Nothing was to rival it for size for 50 years.

Herschel was bettered only by a combination of astronomical passion and vast wealth. Irish landlord William Parsons, Earl of Rosse, wanted the biggest telescope ever built. By 1847, despite the Potato Famine, he had it. 'The Leviathan of Parsonstown' boasted a 20-metre-long tube and a 2-metre mirror of polished metal (mirrors made of silvered glass were not yet available). Hung from massive masonry walls and wooden scaffolding, the telescope could observe stars and planets for only a couple of hours as they passed overhead.

Although it operated for only a few decades, the Leviathan performed magnificently, revealing stars 10 000 times fainter than the eye can see unaided; tiny craters and rills on the surface of the Moon; and new and delicate details on the surfaces of Mars, Jupiter and Saturn. Faintly glowing gas clouds or 'nebulas' were turned into stunning spirals of light.

In time, even bigger telescopes would be built, collecting 20 times more light than the Leviathan. Even more wonders were discovered with them.

Ending the Mess Over Measurement

Gabrielle Mouton

Every time, place and culture devised its own system of weights and measures, based, say, on the length of the king's foot. The confusion complicated trade, even within one nation.

The French Revolution generated today's solution to the tangle, though the basic idea is much older. In 1670, the priest Gabrielle Mouton from Lyons proposed that the unit of length be something everyone could agree on, like a fraction of the circumference of the Earth. People also wanted a simple system, rather than absurd intricacies like '3 feet to the yard, 22 yards to the chain, eight chains to the mile' (to take an English example).

In 1790, with the French Revolution underway, the parliament, urged on by the political leader, Talleyrand, asked the Academy of Sciences to devise something better. The resulting 'metric system' used the ancient Greek word for 'a measure'. The units of, say, mass were all 10 times larger or smaller than each other. A 'metre' was to be exactly one ten-millionth part of the distance from the equator to the pole, calculated by measuring the Dunkirk–Barcelona sector (that took six years). But the surveyors made some errors in those turbulent times; the locals were often suspicious of their activity and got in their way.

In this system, a thousandth part of a cubic metre defined a 'litre'. A litre of pure water weighed a kilogram. Prefixes like *milli-* and *kilo-* came from Greek and Latin words for a thousand; *mega-* and *micro-* meant 'very large' and 'very small'. It was logical and became law in France in 1795. For convenience, standard bars and blocks of platinum were constructed to define metres and kilograms.

The system went in and out of favour for decades. Napoleon banned it for a while, though French conquests spread knowledge of it over much of Europe. France made it compulsory in 1840, and the rest of the world. The USA is the only advanced nation where the metric system is not in everyday use, even though it is legally sanctioned, and used in science and technology.

An international 'Treaty of the Metre' was signed in 1873 and the system has been extended and refined many times since. Nowadays the units of length (and time) are measured against the properties of light waves of a certain precise colour, rather than the size and movement of the Earth. Universally used in science, the metric system has needed new vocabulary; *giga-* (a billion) and *nano-* (a billionth) come from old Greek words for 'giant' and 'dwarf'.

The Carburetor and its Kin

Giovanni Venturi, Henri Pitot

Every early motor car (and some modern ones) had a carburetor, a device that mixed air and fuel to go into the cylinders. This invention, like a number of others, reflects our debt to Giovanni Venturi. The name of this Italian priest and physicist, active at the turn of the nineteenth century, lives on in a simple device with many uses in pumps and pressure gauges. A 'venturi tube' gets narrower and then wider along its length, making fluids like air or water speed up and then slow down again as they flow through.

By Venturi's time, we knew that as fluids flow faster, the pressure within them lessens. So in the narrow middle part of the pipe, the pressure is lower than at either end: this is the 'venturi effect'. If we measure that pressure difference, we can calculate how fast the fluid was moving when it entered: the pressure drop depends on that as well as on the changing diameter. So a 'venturi meter' can tell how fast water is flowing past the hull of a boat, how fast a plane is flying, how fast the wind is blowing (using a 'Dynes anemometer').

If another tube is connected to the central low-pressure region, some other fluid can be drawn in to join the stream. Such a 'venturi pump' can empty water out of a boat simply by the boat's motion, or pull air out of a closed vessel by powering the pump with a stream of water from a tap.

In every petrol engine 'carburetor' air is sent through a venturi tube. The side tube into the venturi connects to a source of petrol, which is sucked into the air stream by the lower pressure (like sucking a milk shake through a straw). What is more, the petrol quickly turns to vapour, since lower air pressure makes liquids more volatile. So the carburetor sends into the engine not air mixed with drops of liquid petrol but air spiced with carbon-containing petrol vapour, which is what the name 'carburetor' refers to.

Venturi did not invent all of this. Some of the applications (as in fly sprays) came later. And years before, the great mathematicians Leonard Euler and Daniel Bernoulli had explained just why the pressure in a liquid or gas drops as the speed of flow increases.

Earlier still, around 1732, French engineer Henri Pitot devised a tube for measuring the speed of air or water flows, using much the same idea. When employed to measure the speed of a plane, such devices are still called Pitot tubes in his honour.

The Quest for the 'Alkali'

Nicolas Leblanc

Late in the eighteenth century, industries making glass, soap, textiles and paper were beginning to boom, driving up demand for raw materials. Chief among these were the alkalis, particularly soda ash (aka washing soda), made by boiling down various seaweeds such as Scottish kelp and 'barilla' from the Canary Islands. Supply was unreliable and quality was variable. More abundant and reliable sources were urgently needed.

The French, at war with much of Europe and unable to import soda ash, were in particular need. In 1783 King Louis XVI had his Academy of Science offer a prize to anyone who could 'make the alkali' (that is, soda ash) by decomposing sea salt by the 'simplest and most economical method'.

The starting point was already known. German doctor Johann Glauber had found it a century before: he heated sulfuric acid with sodium chloride (sea salt) to make hydrochloric acid. This left a residue that he called Glauber's Salt (it was used in medicine); a popular name was 'salt cake'; formally it is sodium sulphate. Various people thought soda ash (sodium carbonate) could be extracted from the salt cake but no one could do it cheaply and easily enough to meet demand.

Nicolas Leblanc, a doctor employed by the Duke of Orleans, came up with his 'process' in 1791. He roasted the salt cake with charcoal (carbon) and chalk (calcium carbonate). Off came carbon dioxide, leaving a mix of the desired sodium carbonate and calcium sulphide waste; the two could be separated by washing. The problem appeared solved. Leblanc's patron built a big plant outside Paris and production began. Demand was high because of war.

However, the outcome was tragic for Leblanc. The plant was soon appropriated by the revolutionaries, the duke beheaded, and Leblanc forced to reveal the secret of his process. Others quickly put it to work. He never received compensation for the takeover of the plant, and by the time he resumed production his competitors were too far ahead. The academy had been dissolved, so he never got the prize. Bankrupt and despairing, he finally shot himself.

His process, too, ultimately came to nothing, even though it was good at making soda and hundreds of plants were built. Environmentally unfriendly, with various nasty wastes, it was ultimately superseded by the Solvay method. At least his descendants finally got some reward. In 1855 Napoleon III made a payment to them in lieu of the prize. A statue of Leblanc now stands in Paris.

Information Outruns the Horse

Claude Chappe

At the end of the eighteenth century, the fastest and most reliable way to send a message further than you could shout was to give it to a man on a horse. Forty-year-old French clergymen Claude Chappe came up with something faster. With the aid of four unemployed brothers, Claude determined to perfect something proposed but never achieved in ancient times, a network of relay stations to transmit messages over long distances in minutes or hours, rather than days.

France being at war, the authorities soon accepted the military value of such a system. The support of a fifth brother, Ignace, a member of the revolutionary parliament, smoothed the way.

The first chain of such stations opened in 1792 from Paris to Lille, about 150 kilometres. The stations, about 15 kilometres apart, were equipped with a tower supporting a jointed crossbar that could be arranged at various angles and patterns to signify letters. The station operators kept watch up and down the line using telescopes, sending on as they arrived. At peak efficiency, Paris to Lille took as little as five minutes.

The number of 'telegraph' lines (the official term, devised by Ignace and meaning 'distant writing') soon spread across France from coast to coast and reached into nearby countries such as Belgium and Italy. A message could pass from Paris to Venice in six hours. Napoleon came to rely on the system to coordinate his realm in peace and war, using mobile stations on carts to bring news from the battlefront. Other nations took up the idea, including Russia, Sweden and England, though in the last poor weather often reduced visibility and slowed down the traffic.

Chappe's system ultimately came under challenge from the newer 'electric telegraph', though one line remained in use in Sweden until the middle of the nineteenth century, and officials at the British Admiralty thought the 'semaphore' they were running from London to Portsmouth worked well enough for them to turn down a proposed electric telegraph in 1819.

Chappe's telegraph has a place in literature. In Alexander Dumas's novel, the Count of Monte Cristo bribes a poorly paid operator to send a false message.

All did not go well for Chappe, despite his founding the world's first telecommunications network. Depressed by illness, and harassed by claims from rivals, he committed suicide in 1805 by throwing himself down a well at his Paris hotel.

Lighting the Streets with Coal

William Murdoch

When Abraham Darby (p. 54) first heated coal in a closed retort in 1709 to make coke for smelting iron ore, he would not have known how much of value was going up the chimney. Coal is a complex mixture of organic compounds. Heating in the absence of air (so that the coal does not burn away) releases many of these, and with care they can be trapped and collected for use.

The first useful find was coal gas—colourless, if smelly. Composed mostly (we now know) of hydrogen, methane and carbon monoxide—all of them flammable and the last poisonous—coal gas had been made and burned in small amounts previously. Not until the 1790s was it made in large amounts for distribution through buried pipelines to light houses and streets.

In 1792 Scottish engineer William Murdoch showed the way. He lit a house in Redruth, Cornwall, with flames from a burning jet of coal gas. Existing domestic lighting came from candles and oil lamps; house interiors were dim at night. Streets were lit by oil lamps, if at all. In the absence of moonlight, careful pedestrians carried their own lamps.

Over coming decades, coal-gas lighting spread to the streets of major towns and the houses of the well-to-do. Gas heating and cooking soon followed. To even out supply and demand, Murdoch devised large expandable storage tanks that he called 'gasometers'. His countrymen like to claim Murdoch as 'the Scot who lit the world'.

Murdoch had once walked 500 kilometres to Birmingham to get a job in the steam engine factory run by James Watt and Matthew Boulton (p. 51). He became a long-serving and much-respected member of the firm, and later a director. While attending to the firm's interests among the Cornish tin miners he rediscovered the illuminating power of coal gas. He lit the Birmingham factory with gas in 1803.

He also pursued the technology of high-pressure steam engines, but Watt and Boulton were not interested, mostly because they thought them unsafe—which was often true—and dissuaded Murdoch from further work.

Like Boulton and Watt, Murdoch was a member of the Lunar Society of Birmingham, sometimes dubbed the 'think tank' of the Industrial Revolution. The society met on nights of the full Moon, so to light the path homewards through dark streets after hours of discussion about the future of industry. After Murdoch's innovation, they could have met on any night.

Off with their Heads!

Joseph Guillotin

Many people have given their names to inventions, but few bring as macabre an image as the French doctor Joseph Guillotin and his brainchild. Strictly speaking, the good doctor (as he was reputed to be, kindly and humane) did not invent this instrument of execution. He merely advocated its use (and improved its operation) to reduce the suffering of those condemned to die, speeding their departure.

The history of what is now called the guillotine goes back many centuries. According to the Chronicles of Hollingshead, something very similar was used to execute one Murcod Branah in Ireland for an undisclosed crime in 1307. A 1577 woodcut shows a wooden platform, the victim lying face downwards on the block; over him a heavy blade is ready to descend. The Halifax Gibbet, a similar device, was last used for execution in 1648. There were numerous similar 'engines' throughout Europe, replacing the axe or the sword.

The machine bears Guillotin's name because, as a member of the new French parliament, he strongly (and effectively) advocated its use as the preferred (indeed the only permitted) method of execution, applicable to all classes and carried out without torture. It would replace much slower and more brutal methods of execution, such as the stake, the garrote and the noose. Punishment for a corporal crime would be 'by decapitation, using a simple machine'.

The first guillotine, reputedly constructed by a German harpsichord-maker at a cost of 960 francs (including a leather bag for the severed head), was tested on sheep and calves and then on human cadavers. A 40-kilogram blade was let fall from a height of 2 metres. Various improvements included a straight rather than a curved blade, and brass specified in key elements.

Highwayman Nicholas Pelletier was the first criminal to feel the lethal blow of 'Madame Guillotine', on 25 April 1792. A year later, the mass execution of aristocrats (and royalty) at the hands of the revolutionaries began, with 40 000 of the condemned rolling through the streets in tumbrels for their appointment with the 'national barber'. At the tailend of the 'Terror', the leading revolutionaries came to the same end.

Further improved, the guillotine was used in public until 1939 and, hidden from public gaze, until 1977, when the death penalty was abolished in France. The name is now given to a long pivoted blade used to trim paper.

Cotton Seeds and the Civil War

Eli Whitney

Eli Whitney's 'cotton gin' changed history. It transformed productivity and profits on the cotton plantations in the American south. Escalating production necessitated the continued (and growing) use of slaves to work the fields. The American Civil War followed almost inevitably.

To allow spinning and weaving, cotton growers had to separate the fibres from the small sticky cottonseeds. These could be pulled out by hand, but a worker would struggle to clean half a kilogram of cotton in a day. Existing machines to do the job, such as the East Indian charka, which pulled the cotton between rollers, did not work with the short-fibre cotton grown in the south away from the coast.

Eli Whitney had been fascinated by machinery since childhood. A law graduate from Yale and in need of money to pay off debts, he became a tutor on the Georgia plantation of Catherine Greene. Seeing the need for a machine to clean cotton, he quickly devised his 'gin' (short for 'engine', meaning a mechanical device). Cotton fibres, snagged on a set of rotating spikes or hooks, were pulled through narrow slots, leaving the seeds behind.

It was simple and very effective. Cleaning rates rose 50-fold, costs fell dramatically, cotton production boomed, doubling every decade. So did the demand for slave labour. By the mid-nineteenth century, American cotton would make up 75 per cent of the world supply, and more than half the nation's exports.

Whitney patented the gin in 1794. According to legend, the concept had come from seeing a cat claw at a chicken through the slatted walls of its coop, grabbing nothing but feathers. His employer may or may not have had the original idea but she certainly gave Whitney financial and moral support. He needed it: Whitney and his business partners struggled to protect their patent, which was routinely infringed, and to make any money from the innovation.

In one way, the cotton gin hurt the South. Cotton-growing became so profitable that it swallowed up nearly all investment, leaving little for manufacturing or the cities. Nearly all new manufacturing industry moved north, boosting populations there. The South was left badly under-resourced when the Civil War came, and perhaps ensuring its defeat. Ironically, Whitney would make more money from his methods for the mass production of guns (p. 101), such as those used in the war, than he ever did from the cotton gin.

The Battle Against Smallpox

Edward Jenner

In the eighteenth century, smallpox was still a dreaded disease, able to blind, maim, disfigure or even kill those who contracted it. No one knew what caused it, and treatments were 'rule of thumb' and generally ineffectual. The toll was vastly greater among the indigenous populations of Africa, America and Oceania, where smallpox was unknown until the coming of European invaders.

Since the start of the century, 'inoculation' had been practised in Europe as a way of conferring some sort of immunity to the disease and preventing its spread. Smallpox sufferers rarely got the disease twice, suggesting that one infection generated resistance to a second attack. In inoculation, pus from a pustule on a smallpox sufferer was rubbed into a small incision on the skin. The patient was thereby given a mild dose of the disease from which they would hopefully recover and be immune thereafter. The drawback was that until the disease had run its course, the patients were infectious and could pass the scourge on to others.

English country doctor Edward Jenner, who practised in Gloucestershire, put forward a better technique. He drew on a local folk tradition that milkmaids had attractive, unblemished skin because they had contracted a mild infection called cowpox from their daily work and this protected them from the much more serious smallpox. In now appears that similar contact with domesticated animals gave us at least partial immunity to a number of diseases that had cut a swathe through 'new world' populations without such close contact with animals.

In 1796 he tested the idea, using the method of inoculation, but with pus from a cowpox pustule. The patient, eight-year-old James Phipps, quickly recovered. Six weeks later Jenner gave him the standard smallpox inoculation and the boy was completely unaffected. The cowpox had indeed conferred immunity against smallpox.

Jenner called his new method 'vaccination', from the Latin *vacca* meaning 'cow'. It was the start of a new era in public medicine, and was to lead 200 years later to the total global eradication of smallpox and the near-disappearance of polio. From the work of Louis Pasteur and others in the next century (p. 143), ways were found to confer immunity to many other deadly diseases.

Daredevils with Parachutes

Jean-Pierre Blanchard

Many people contributed to the invention of the parachute. Leonardo da Vinci (p. 12) is often credited with the idea. Certainly he sketched one around 1485. Others before him left drawings of similar devices. Medieval documents tell of adventurous folk jumping and using something to arrest their fall, such as a loose cloak stiffened with wooden rods. Some survived, some did not.

The story picks up pace with the growing popularity of ballooning (p. 57). Frenchman Sebastien Lenormand of France coined the name, meaning 'to protect against a fall'. His creations, like earlier designs, used linen stretched over a frame. Two years later, daredevil Jean-Pierre Blanchard tested one using his dog. The dog survived, as did Blanchard himself in 1793 when his hot-air balloon caught fire and he had to 'bail out'.

Something more like the modern parachute appeared as inventors sought more compact and easy-to-use designs. Blanchard led the way and by the late 1790s he was using folded silk, which was both light and strong, but without a frame. In 1797 his compatriot, André Garnerin, made the first jump with such a parachute. Perhaps this is the key date.

More than a century later, the parachute found a ready market with the arrival of powered flight. In Missouri in 1912, eight years after the Wright brothers (p. 172), US Army Captain Albert Berry entered history as the first person to parachute from a moving plane. Gender equity was restored in 1913, when crowds in Los Angeles saw Georgina Broadwick repeat the feat. The next year, Charles Broadwick patented the 'backpack' parachute, much as is used today.

In World War I, parachutes allowed crews to escape from tethered observation balloons. Hydrogen-filled, these were tempting targets for the early fighter planes. But the pilots were not issued with parachutes, partly because the mass-produced canvas parachutes were too heavy for the primitive planes to carry, and partly from a fear that having such an easy means of escape from battle would promote cowardice among inexperienced flyers.

Not until 1918 were German fighter pilots given parachutes. These were deployed by 'static lines' fixed to the plane; these caused problems when the plane was spinning. In 1919 US inventor Leslie Irvin introduced the 'rip cord', letting the pilot activate the 'chute when he was clear of the plane. The first life so saved was reportedly that of William O'Connor in 1920.

How We Got
the Soda Fountain

Joseph Priestley, Samuel Fahnestock

For nearly a century, 'soda fountains' in 'drugstores' (pharmacies) were common meeting places for people, especially the young, in middle America. The rest of the world saw it in the movies. Times have changed, but the story is still worth telling.

In 1772 English chemist and radical cleric Joseph Priestley reported some experiments with the newly identified 'fixed air', the gas we now call carbon dioxide. He had dropped sulfuric acid onto limestone and induced some of the resulting gas to dissolve in a pan of water nearby. He found the water now had a pleasant, slightly sour taste and an effervescent sparkle, like naturally occurring mineral waters.

In one version of the story, the first source of the gas was a local Leeds Brewery, since carbon dioxide is produced plentifully during fermentation. Quite independently, the Swedish chemist Torbern Bergmann had made the same discovery. The two had invented 'soda water', though the name was not used until 1798. 'Soda' (sodium carbonate) released carbon dioxide when heated or treated with acid, and became the preferred source of the gas.

Since the drinking of 'carbonated waters', either natural or artificial, was considered healthy, pharmacists or drugstores soon took an interest in selling such drinks, particularly in the USA. The 'soda fountain' for mass dispensing was invented by Samuel Fahnestock in 1819 and John Matthews' 'apparatus for charging water with carbon dioxide gas' made the process much faster and cheaper around 1832. Around 1861 flavoured fizzy drinks, already commonly called 'soft drinks', were first dubbed 'pop' (from the bubbles); 'ice cream sodas' appeared around 1874.

Every pharmacy developed its own specialties, adding herbs and plant extracts to increase flavour and (perhaps) the health benefits: birch bark, dandelions, sarsaparilla and various fruits. Memorable outcomes of this trend were Dr Pepper (invented by Charles Aderton, Waco, Texas in 1885) and Coca-Cola (Dr John Pemberton, Georgia, 1886). 'Coke' contained extracts from the caffeine-rich cola nut and (until 1905), traces of cocaine, as it was marketed as a tonic.

Sales of drinks from the Coca-Cola Company are now a billion a day worldwide, but only nine people bought the drink on its first day of sale in a drugstore, and the company lost money in its first year. The company began to distribute the (still secret) syrup to independent bottlers, sales of syrup increased by 4000 per cent from 1890 to 1900.

Better Paper, Faster Printing

Nicholas Robert, Friedrich Koenig, William Bullock

Paper, like many seminal inventions (including gunpowder, the compass and pasta) reached Europe from China, helped by exploration, trade, war and religious conquest. Initially all paper was made from linen and cotton rags, as the finest quality papers are today. Fabric scraps were broken down into fibres in water and treated with glue-like 'size' to hold the fibres together. A filler such as starch or chalk was added to smooth and whiten the surface. The fibres, once spread on a sieve-like frame that allowed water to be squeezed out, were pressed into a sheet and dried. It was a slow process; paper remained expensive.

With the spread of literacy and the rise of magazines and newspapers (*The Times* was first published in London in 1788), the demand for paper rose rapidly. Papermakers turned to other sources for fibres. In 1720 versatile French scientist René Réaumur noted the way certain wasps created paper for their nests from plant materials. This stimulated experiments, especially in Germany and France, with papers made from materials as diverse as hemp, straw, tulip leaves, potatoes, stinging nettles and cabbage stalks.

Wood was harder to turn into paper. The timber had to be ground into pulp, and not until 1840 was a patent secured (by the German weaver Friedrich Keller) for a wood-grinding machine for the purpose. Early pulp paper was enriched with cotton or linen fibres for strength and quality, and not until the 1880s was all-wood-pulp paper judged an adequate substitute for cotton/linen paper. Early 'newsprint' quickly went yellow and brittle due to the presence of lignin from the wood, giving rise to the derogatory term 'yellow press'.

As to the papermaking itself, the first steps toward mechanised production to increase speed and reduce costs were taken around 1798 by a young French student, Nicholas Robert. His experimental machine reproduced the various steps of the manual papermaking process but made a continuous roll of paper (rather than separate sheets) by placing the fibres on an endless wire mesh with squeeze rollers on one end.

This became known as the Fourdrinier machine, from the involvement of the London stationers the Fourdrinier brothers, who perfected and marketed it. They had craftsmen build a working prototype machine based on Robert's original plans and models, and by 1803 were able to make paper of reasonable quality.

The pair spent a considerable amount of money (several million dollars in today's terms) in developing the machine, but like the original inventor Nicholas Robert they made no real profit from the invention. The rest of us have certainly benefited, as cheaper paper permitted cheaper and more accessible printed materials. The continuous rolls of paper from a Fourdrinier machine were ideally suited for mating with the rotary printing press once that was invented.

At the time, printing, too, was still a one-page-at-a-time process, much as it had been for Gutenberg and Caxton. Ink the bed of type (lines of single letters held in a wooden frame), lay on a page of paper, lower the press to make the impression, raise it again, peel off the printed page...and repeat.

The move forward was driven by the new availability of the steam engine as a continuous source of power, and by the inventive mind of German printer Friedrich Koenig and his chief mechanic Friedrich Bauer. Landing in London early in the nineteenth century, they secured financial backing for their new machine, which mechanised much of the printing process. Rotating cylinders inked the flat type bed and pressed the paper onto it, though the paper was still placed and retrieved by hand.

Printing speeds went up fourfold, from 300 to 1200 pages per hour, and the results were impressive enough to convince the owner of The Times to have two such presses built. The Times went to press mechanically for the first time on 29 November 1814. All the other major newspapers quickly followed.

But the job of mechanised printing was only half done. The final step, the rotary printing press, was taken in 1863 by American inventor William Bullock, improving greatly on a machine built and patented 20 years earlier by his countryman Richard Hoe. Bullock made a cast of the type bed and curved it to fit a cylinder. A continuous roll of paper could be fed between one roller carrying the type and one providing pressure, with a third roller inking the typebed between impressions. Printing speeds soared (up to 12 000 pages per hour); the machines could print both sides at once, and cut and fold the paper. Printing technology is fundamentally the same today, though the nature of the printing plates has changed.

In a sad and bizarre accident, Bullock's leg was shattered when he become entangled in one of his machines, and he died after gangrene invaded the leg.

Machines
that Make Machines

Henry Maudslay

Henry Maudslay is not a name many would recognise, but he has a central place in the development of machine tools, vital for all manufacturing industries. He worked for a time for Joseph Bramah, inventor of the safety lock and the hydraulic press (p. 59); his protégés included Joseph Whitworth of the standard screw thread (p. 103) and James Nasmyth of the steam hammer (p. 98).

At 12 years old Maudslay was a 'powder monkey', carrying gunpowder at the Woolwich Arsenal, where his father was a wheelwright, moving on to the woodshop and the smithy. His growing skill and reputation saw him recruited by Bramah when still in his teens. Maudslay married Bramah's housemaid, and six years later (three years as foreman), with three children to support, a request for a modest pay rise was refused. Maudslay walked out to start his own works in Oxford Street London and later in Lambeth.

Maudslay's major contribution to the growing power and precision of metalworking and manufacture was the first modern lathe, able to precisely cut the threads on a screw. A Frenchman had first built such a machine from wood in 1568, but that would never match the demands of the looming industrial age. Metal-cutting lathes today look very much like Maudslay's 1800 machine, although American inventors added their own ideas. Big lathes are now controlled by microprocessors.

Cutting with precision required accurate measurement. The best available instrument was a micrometer (p. 27) devised by James Watt, a master instrument-maker before he turned to steam engines. Turning a screw caused a plate to move forward by distances measured to a hundredth of an inch. Maudslay took that as a starting point and by 1905 had built his grandly named 'Lord Chancellor' micrometer. With the screw holding 100 turns to the inch, and a measuring disc dividing each turn into 100 parts, the limit of measurement became a ten-thousandth of an inch.

Maudslay's new lathe arrived as the patents on Watt's steam engines (p. 51) expired and engineers all over the country began to design and build their own. One of Maudslay's new machines gained major navy contracts, and was used for punching holes in metal plates for riveting into boilers and tanks. This had previously been done by hand. Maudslay's factory remained one of the most important precision machine shops throughout the nineteenth century, closing only in 1904. But Maudslay had died in 1831, aged 60.

The Age of Electricity

Alessandro Volta, Humphry Davy

Throughout the late eighteenth century, the courts and drawing rooms of Europe were entertained by tricks played with 'electrical machines', generating what we would now call 'static electricity' through friction. Participants could give each other electric shocks while kissing, set fire to flammable liquids with sparks and watch paper and feathers flying around through electrical attraction.

It was all good fun, and the early scientists gained some insight into the nature of the phenomenon. But there seemed little useful that could be done with such charges, since they could not be sustained. One spark, one shock and they were used up. A continuous flow of electric charge, an 'electric current', might be useful but no-one knew how to produce one.

A device to produce such a current was found in a curious way. The Italian anatomist Luigi Galvani found that pieces of two different metals, such as copper and iron, in contact with animal body fluids, could make things happen, such as sudden movement in apparently lifeless tissues, such as frogs' legs. Galvani put it down to 'animal electricity', some 'life force', but a few years later his countryman Alessandro Volta hit on the real explanation.

It was all just a matter of chemistry. Two dissimilar metals on either side of a piece of paper or cloth soaked in a salty or acid solution made a 'cell' (probably meaning a small room, or the box-like structures seen in plants with a microscope). Connect the metals together with a wire and a current would flow, sustained by a chemical reaction, at least until the chemicals were used up. Connect a number of cells to increase to increase the current and you had a 'voltaic pile', or, in words we use today, a 'battery'. The analogy was with an artillery battery, a number of guns side-by-side firing at once.

The New Wonder

Volta's battery was soon the rage of scientific Europe, and researchers rushed about to find new things to do with the current. The inventor was summoned to demonstrate the device before Napoleon, who had a good feel for the importance of such discoveries.

Electricity proved a powerful and immensely versatile technology, providing new forms of lighting, powering electromagnets, the electric telegraph and the telephone, running electric motors to provide power for transport and industry and ultimately giving rise to 'electronics') to create a communications revolution.

The 'Age of Electricity', which dawned in 1800 with Volta's invention, persists to the present time, and will into the future. Electricity has become essential for everything we do.

In a voltaic battery a chemical reaction generates an electric current. England's Humphry Davy, among others, found that the reverse was also true. Electrical currents could drive chemical reactions. Chemical compounds could be broken into their elements or reformed into other compounds ('electrolysis'); aluminium, magnesium and chlorine are prepared in this way today, as is 'caustic soda' (sodium hydroxide).

Davy and others also found electricity could deposit a film of metal (such as silver) from solution onto objects like knives and forks ('electroplating'). This was dubbed 'galvanising' in honour of Luigi Galvani.

Davy was one of the electrical pioneers. As well as his work on electrolysis, he ran a strong current between two pieces of carbon that were almost in contact. The resulting brilliant white light (the 'electric arc') was lighting by electricity 70 years before Edison and Swan's light globe.

Looking for Better Batteries

Over the two centuries since Volta's invention, all sorts of combinations of chemicals have been tried in search of batteries with better performance, greater reliability and more energy released from a smaller volume or weight. The pairing of carbon with zinc is found in the ubiquitous 'dry cells' (AA, AAA and other sizes) that power all sorts of portable appliances. More advanced types, such as the 'alkaline' batteries, have extra chemical wrinkles to make the current last longer.

Today, many batteries use the metal lithium, which make them not only light and compact (a major reason why mobile phones and laptop computers get smaller every year), but also rechargeable; energy can be stored in the battery by driving the chemical reaction in reverse.

The most common rechargeable battery is still the venerable lead/acid type invented in the nineteenth century and found in every motorcar—cheap, reliable but heavy and not really environmentally friendly.

More exotic rechargeable batteries have been proposed and developed, but are not yet in widespread use, for reasons of cost or challenging operating conditions (such as running at high temperature).

Yet the demand for new types of battery is high, for example to store renewable energy such as wind and solar power and make it competitive. More than 200 years after Volta, we are in dire need of a new breakthrough in battery technology as big as the one he gave us.

Machines that Need No Watching

Joseph Jacquard, Charles Babbage

Devices able to carry out complex tasks without continual human guidance, such as computers, are not merely twentieth-century marvels. In the previous century, the Jacquard Loom could do its job unsupervised, and Charles Babbage's 'analytical engines' were an early vision of calculations controlled by a 'program'.

Through hard times and the Revolution, French silk-weaver Joseph Jacquard kept trying to automate weaving. An endless loop of punched cards controlled a loom he built in 1801. The holes in the cards determined which longitudinal ('warp') threads were depressed during each pass of the shuttle; complex patterns could be woven without the weaver intervening.

At first Jacquard's looms attracted the ire of weavers, who feared unemployment, like the Boston tailors in sewing-machine inventor Elias Howe's time (p. 108). But with government support their numbers soared. Eleven thousand such looms were in use by 1812. A threshold had been crossed.

Charles Babbage, son of a banker, was a brilliant mathematician and holder of the Lucasian Chair at the University of Cambridge previously occupied by Isaac Newton. His many achievements include the establishment of the British Association for the Advancement of Science, and advances in making and breaking codes. He is mostly remembered today for proposing (and failing to complete) the 'difference engine' and the 'analytical engine', the first attempts at calculations without continual human involvement.

To minimise error in the laborious calculations of mathematical tables, including those needed for navigation by ships (p. 50), the steps in the calculations would be controlled by punched cards, as in Jacquard looms, rather than by a human operator prone to fatigue. In the days before electronics or even electricity, the 'engines' would be purely mechanical, with thousands of gear wheels and axles, and likely to weigh tonnes when completed.

Such an enterprise was unachievable without government support. This, Babbage initially secured but later lost as the usefulness of his project came under fire, and as he encountered limitations imposed by the technology of the day. He was also constantly at odds with the craftsmen he employed from the 1830s to build the engines. Only fragments were ever completed, and after a lifetime of labour, Babbage died embittered by failure.

Babbage's vision was not achieved for a century, until electronics emerged as the technology for the task (p. 224). In honour of Babbage, some computer programs, such as those used to search the Internet, are called 'engines'.

Steam across the Water

John Fitch, Robert Fulton, Henry Bell

Very early in the development of steam engines, inventive spirits like Denis Papin (p. 34) dreamed of (and experimented with) boats driven by steam rather than by the wind or human effort. But many decades passed before those visions became reality.

Jonathan Hulls took out patents on a steamboat in 1736, but it was to be driven by a Newcomen engine (p. 37)—heavy and inefficient—and would never be a success. In 1763 William Henry in Pennsylvania put a Watt steam engine (p. 51) into a boat, but it sank. Two decades later in France, a steam-powered paddle-wheeler managed 15 minutes against the current on the River Saône, but lacked the endurance for longer trips.

Developments elsewhere included a boat driven by a steam-powered water-jet and able to do 6 kilometres per hour, but all these were too slow, or too expensive to run, to be attractive. For example, American John Fitch successfully trialled his first steamboat in 1787; a later version carried passengers. Fitch tried a number of designs and solved many technical challenges; one of his boats could travel at 13 kilometres per hour. Yet he could never convince skeptics that steamboats would pay.

The nineteenth century arrived before real success could be claimed. In Scotland in 1802 Lord Dundas launched the *Charlotte Dundee,* a double-hulled paddle-wheeler with an improved engine by William Symington. This pulled two 70-tonne barges 30 kilometres along the Forth and Clyde Canal to Glasgow.

Soon after, success came to American Robert Fulton, whose countrymen call him 'the father of the steamboat'. Inspired by news of the *Charlotte Dundee,* Fulton ran trials on the River Seine, where he was hoping to get French support for his submarine *Nautilus* (p. 164). Once home, he imported a Boulton–Watt steam engine and built a boat to take it. In 1807 the *Claremont* began a regular passenger service between New York and Albany, 250 kilometres up the Hudson River, taking 30 hours for the trip. Within a few years, steamers were running on the St Lawrence River in Canada and would soon appear on other rivers and lakes, including that most famous venue for paddle-wheelers, the Mississippi.

Back in Scotland, Fulton's ideas inspired Henry Bell, who launched his *Comet* in 1812 onto the Clyde between Glasgow and Greenock. Inside a decade, dozens of smoking steamers were plying the rivers, lochs and canals of Scotland, carrying cargo and occasionally passengers. The age of steamboats had come.

Making Food Keep

Nicholas Appert, Peter Durand

Over the millennia, many ways have been found to delay food 'going off', so it can be stored for long periods: drying, smoking, salting, pickling, impregnating with sugar, the use of spices or certain chemicals and chilling in ice or snow. All of these methods work for the same reason. They prevent, or at least slow, the activity of 'microorganisms' (yeasts, fungi, bacteria) which, as the French scientist Louis Pasteur would later show, cause the chemical changes we call 'going bad' in the simple pursuit of the energy they need to multiply.

By the early nineteenth century, it was well known that if a meat broth, for example, was first boiled and then sealed from contact with the air, it would 'keep' (and be safe to eat) for several weeks. Again it was Louis Pasteur who showed why: the boiling killed the microorganisms, and isolation of the food prevented further contamination.

A major step forward in food preservation, and indeed in convenience, came early in the nineteenth century with the first sealing of cooked foods into metal containers. A little earlier, in 1807, French pastrycook Nicholas Appert pioneered storing food in glass jars with cork stoppers, so securing a prize offered by Napoleon, who was concerned about providing safe food to his troops in far-flung battlefields.

London merchant Peter Durand took the idea further around 1810, using cans of iron coated with a thin layer of tin to minimise rusting (hence 'tin cans' or simply 'tins'). Tinned food reached the London shops around 1830, though for decades it was used mostly by the military and on voyages of exploration. Indeed, inadequate preparation of canned foods by an unscrupulous supplier helped precipitate the disaster that overtook the Franklin expedition in search of the Northwest Passage from the Atlantic to the Pacific in the 1840s.

Getting the food out of the can and into the pan was a separate challenge. Early tins were made of thick, heavy iron, weighing half a kilogram even when empty. You needed a hammer and chisel to open them, which limited their appeal to the householder. By the 1860s, tins of lighter, thinner steel with a rim around the top made possible the can opener, beginning with the bullhead type around 1885, with a lever-like body and a point to pierce the can. Much ingenuity has been expended on alternative designs in the century or more since.

Trains Running on Iron Rails

Richard Trevithick, George Stephenson

Using rails to carry and guide vehicles is an old idea. While the steam railway was a nineteenth-century invention, miners in Transylvania had used wagons running on wooden rails to carry out ore since the sixteenth century. These were much easier to push and pull than wagons running over rough ground, and the practice soon spread. Coalmines in England, Germany and other countries, increasingly important as the Industrial Revolution spread, were equipped with wooden tracks from the workface to the pithead. If there was room, horses were used to pull the wagons, and a hardy breed of 'pit ponies' developed. Otherwise the labour was human, unless gravity was able to help.

Increasing use wore out the wooden rails; their life was extended by covering them with removable strips of hardwood, later strips of iron, and finally by making the rails themselves of cast iron, which was strong and hard wearing, though brittle.

What we now call railways became possible in the early nineteenth century. Improvements in power and efficiency allowed a steam engine mounted on wheels to pull more than its own weight. In 1804 the pioneering engineer Richard Trevithick first used a steam engine to pull trucks in a mine in South Wales. But the engines were still so heavy they often cracked the cast-iron rails. So mine owners mostly shunned the 'iron horse' and stuck to the real ones. The notion of using steam engines and rail tracks to carry goods (and perhaps even people) over long distances seemed fanciful.

As always, some visionaries could see the future of the new technology. Chief among these was Englishman George Stephenson. Coming from a mining family, he knew how valuable a steam 'locomotive' could be; skilled in mechanical matters since his youth and familiar with James Watt's steam engines he thought he knew how to build one. The 'Blucher', which he trialled successfully in 1814, was the first practical locomotive, hauling coal at the Killingworth colliery in Northumberland. He soon improved it by use of the 'steam blast', directing steam exhausted from the cylinders up the boiler chimney. This made the fire hotter, increasing power and efficiency, and getting more work from each tonne of coal.

There were bigger things ahead for George Stephenson, and for the 'railways' (p. 92).

Reading the Information in Light

Joseph Fraunhofer

The purpose of some technologies is to increase knowledge, rather than to meet some human need. Telescopes and microscopes, barometers and thermometers fall into this class, as does the 'spectroscope'. Spectroscopy means roughly 'seeing the colours (of light)', and it is a powerful tool for science today. Its origins lie with the great English scientist Isaac Newton. In 1672 he had reported that passing sunlight through a glass prism broke it up into a continuous spectrum of colours—the colours of the rainbow from red to violet.

Around 1814 the methodical German, Joseph Fraunhofer, got busy. Fraunhofer made optical equipment, and to begin with broke sunlight into colours just to prove how good his prisms were. He charted with care the many fine black spectral lines that others had found when analysing sunlight, increasing to over 500 the number of what were soon called 'Fraunhofer lines'. He called the machine that made such observations a spectroscope. About 1821 he invented the 'diffraction grating' to speed the work. This consisted of hundreds of parallel lines scratched very close together on a piece of glass. It spread the spectrum out much more than a prism could, revealing much more detail, and made a spectroscope containing such a grating a very powerful and precise instrument.

Intense study by Fraunhofer and others soon revealed a fundamental truth. Analysing light from glowing gases reveals patterns of bright and dark lines that identify the elements and compounds that gas contains. Each spectrum is like a signature or a fingerprint, unique to the chemicals involved. As a result, we soon found that the stars contain the same elements as the Sun, elements known on Earth. This confirmed that all parts of the universe operate on the same laws of physics and chemistry. New elements previously unknown—including helium, found in the Sun before it was known on Earth—were identified from their spectra.

Subtle changes in the spectra of certain stars showed they were in fact double or triple stars, revolving around each other so closely that no telescope could separate them. In time, the discovery of the phenomenon of 'red shift' in light coming from huge objects called 'galaxies' in deep space would prove that the universe was expanding and indicate its vastness. It is not surprising that the great biologist Alfred Wallace would later include spectrum analysis in his list of the 14 greatest inventions of the nineteenth century (p. 167).

The First Photographs

Joseph Niépce, Louis Daguerre, William Fox Talbot, Frederick Archer, George Eastman

'Writing/drawing with light' (aka photography) is one of the great nineteenth-century inventions, but its origins lie much further back. As early as 1727, German doctor Johann Schulze found that certain compounds (salts) of silver, like silver nitrate, normally pale in colour, went dark when light fell on them. Fifty years later, Swedish chemist Karl Scheele (who discovered chlorine) found that silver chloride darkened under the influence of light because it decomposed into silver and free chlorine.

Thomas Wedgwood, of the famous pottery family, and leading scientist Humphry Davy experimented with these chemicals, and found they could make silhouettes of leaves and other objects by placing them on paper covered with silver salt under sunlight. But they could not make the image permanent; undarkened areas soon went black and the image vanished.

Meanwhile something apparently unrelated was going on. The familiar term 'camera' is Latin for 'room', and goes back to the phenomenon of the 'camera obscura' ('hidden room'). Light from a brightly lit scene let through a small hole in one wall of a darkened room threw a faint but distinct image of the scene upside down on the opposite wall. Since the seventeenth century artists had used miniature versions of this, a small box on legs with a lens in front and an opaque glass screen at the back to view the image. It was a useful tool to frame a scene for painting.

Someone needed to make a leap of imagination here, to bring together the silver salts and the camera obscura, to capture, and maybe preserve, the image on the opaque glass by substituting a piece of glass covered in light-sensitive silver chemicals. That someone proved to be Joseph Niépce, a French amateur scientist. In 1816 he first made images in this way, and even partially 'fixed' them (that is, they did not fade), but they were 'negatives', with black where white should be; he could not make 'positives'. The images also had poor definition and contrast. But it was a start.

More than a decade later, Niépce was still struggling to find the right combination of chemicals to develop and fix his images. He hoped bitumen might provide an answer. Around 1929 he linked up with artist Louis Daguerre but died a few years later. Daguerre continued with some new ideas of his own involving silver plates, iodine, mercury and common salt. Making his 'daguerreotypes' required a tray full of chemicals, but the images were bright and clear. After some improvements in technique, an image could be made in a short enough time for photographic portraits to be feasible.

Lots of people now had ideas on how best to proceed, especially in terms of the chemistry. English astronomer John Herschel proposed new chemicals for fixing the image, and coined the words 'positive', 'negative' and 'photography'. English archaeologist William Fox Talbot found out how to make many prints from one negative, impossible to do with daguerreotypes.

Around 1850 Fredrick Archer of England developed a process for sensitive 'wet plates'. These gave good results but needed to be prepared immediately before exposure and developed straight after. This could be done in the increasing numbers of photographic studios but was difficult out of doors. By the 1870s, photographers could buy 'dry plates', glass rectangles covered with gelatin to hold the light-sensitive chemicals. These needed much longer exposures than the wet plates but worked with landscapes and were much more convenient to use.

Photography as we know it today, done by amateurs and enthusiasts without complex equipment or poisonous chemicals, dates from the 1880s, and the key name is American George Eastman. The new material 'celluloid' (p. 118), available in strong, flexible sheets, was now in production, and Eastman began to market such sheets coated with light-sensitive 'emulsion' for printing photographs.

Once made into rolls, these 'films' fitted into something very new, an easily carried camera called the Kodak. Eastman once said the name had no meaning; its appeal lay in its sound, like a camera shutter opening and closing. First marketed in 1888, the Kodak was sold already loaded with enough film for 100 pictures. When these had been exposed, the camera was sent back to the factory where the pictures were developed and printed, and the camera reloaded.

Other developments followed, perhaps inevitably: films able to be loaded into a camera in daylight so users could do it themselves; moving picture 'films' (p. 158), made using very long strips of the same product; films producing coloured images; safety films that did not burn; and ever simpler and more powerful cameras. This process of development continued until the virtual obliteration of chemically based photography by digital technology and electronics more than 100 years later (p. 279). Kodak stopped making film cameras in 2004.

Noises in the Chest

René Laennec

The name 'stethoscope' comes from Greek words meaning 'looking at the chest', but the organs of 'looking' in this case are the ears. The name was given by its inventor, French doctor René Laennec. Various myths surround the discovery: that Laennec could not hear a patient's heart because street urchins were making a racket outside, and/or that the urchins were scraping on one end of a wooden fence and listening to the sound at the other end. Such details were reported by a former student.

Laennec gave his version of events in an article in 1821. His patient that day was a young woman with a diseased heart, but tapping on her chest provided no information due to, as he put it, 'the great degree of fatness'. Her age and sex meant he could not put his ear directly to her chest.

Suddenly he recalled how the scratching of a pin on one end of a piece of wood could be clearly heard by placing an ear to the other end (remember the urchins?). Laennec did not immediately search around for a piece of wood, but rather made a tube from 20 sheets of paper. He placed one end of the tube on the fat girl's chest, the other to his ear. 'J'entendre!' ('I hear!') he reputedly cried, as the sounds of her ailing heart reached him with unprecedented clarity.

Laennec treated patients with heart and lung diseases at the Necker Hospital in Paris, some with TB (or 'consumption', as it was then called). With his new device, he was soon able to develop new methods of diagnosis, 'a set of new signs of diseases of the chest, for the most part, certain, simple and prominent'. It was a major step forward in medical practice.

Laennec experimented with various materials, seeking the best transmission of sound. Surprisingly, solid materials like glass, metals and wood did not work as well as the column of air in his paper tube, though he quickly moved to wooden tubes 30 centimetres long, turned (reportedly) on his own lathe, with one end flared out into a bell. Some decades later the availability of flexible rubber tubes allowed the development of a 'two-eared' stethoscope, much as we have today.

Laennec himself died of consumption aged only 45, a decade after his great invention and six weeks after returning to his native Brittany.

Magnets and Motors

Hans Ørsted, Michael Faraday

In 1820 Danish physicist Hans Ørsted did a simple experiment and made a striking discovery. He brought a compass needle (a very small magnet) near a wire carrying an electric current from a battery (p. 78). The needle moved: the wire was now acting like a magnet. Electricity and magnetism, both long studied but thought unrelated, were closely linked.

Within a few years, England's William Sturgeon and Joseph Henry in America, among others, found that wrapping many turns of insulated wire into a coil greatly multiplied the strength of the magnetic attraction and produced a north and a south pole like a bar magnet. This 'electromagnet' was even stronger if it had a core of iron or steel. In time electromagnets would be made able to lift tonnes of iron and steel in factories and scrap-yards.

At the other end of the scale, a compass needle hanging in the middle of a coil could indicate even very small currents flowing in the coil by shifting one way or the other. In the soon-to-be invented 'electric telegraph' (p. 97), the flicks of the needle would represent pulses of current grouped according to a code, allowing messages to be sent over hundreds of kilometres.

Pondering Ørsted's discovery, English physicist Michael Faraday at London's Royal Institution noted that the shift in the compass needle when the current flowed indicated that the newly created magnetic field ran around the wire in a circle. Could the effect be used to push a current-carrying wire round and round a magnet, as a sort of 'electric motor'? He made one in 1821. It was not very practical, needing the wire to drag through a bowl of mercury, but it worked. Unfortunately Faraday published this work without due acknowledgement to colleagues. The resulting fuss caused him to bypass work on electricity and magnetism for 10 years.

By the time he came back to the field in 1831, various combinations of magnets (including electromagnets), and coils of current-carrying wire were being tried, and the modern electric motor slowly evolved. These would ultimately find an immense array of uses, driving machinery in factories, appliances in the home, power tools in the workshop, electric trains and trams (and electric cars in our own time). Similar principles would soon be used to generate electric currents (p. 95), hastening the coming of the 'electric age'.

The Taming of Rubber

Charles McIntosh, Thomas Hancock, Charles Goodyear

A raincoat is often called a 'mac' thanks to Charles McIntosh, Scottish chemist and inventor of the rubberised fabric originally used in waterproof garments. The 'macintosh' got its start in experiments with 'naphtha', a liquid waste from the heating of coal in enclosed containers to make coke and coal gas (p. 69).

McIntosh found that naphtha dissolved 'india rubber', made from the sap of some South American trees. He spread the solution between two pieces of fabric. On drying, the pieces were bound together by a layer of rubber, becoming waterproof. McIntosh received a patent in 1823 and opened a factory in Manchester (centre of the cotton trade) soon after. That factory remained operational for more than 150 years. Earlier, in 1820, McIntosh's inventive partner Thomas Hancock patented 'elastic'—strips of rubber attached to cloth. Elasticised female undergarments reached the market in 1830, replacing wire and whalebone 'stays' and definitely increasing comfort; elastic-sided boots arrived in 1837.

Rubber remained troublesome: it was hard when cold and sticky when hot. American hardware merchant Charles Goodyear wanted to make rubber more stable and therefore more useful. He experimented for a decade with combinations of chemicals, but the answer came accidentally when he dropped a mixture of raw rubber and sulfur onto a hot stove. When the smoke cleared, he found the mixture had charred but not melted, meaning that 'vulcanised' rubber was heat-resistant.

By 1841 Goodyear could make thin, uniform sheets of rubber that stayed firm and elastic hot or cold. With little interest in his invention, he tried to sell the idea to McIntosh and Hancock, but Hancock quickly worked out the process for himself and secured a British patent before Goodyear did. Another major discovery was by Alexander Parkes (p. 118): rubber could be 'cold-vulcanised' by dipping in a solution of sulfur chloride. This could make very thin rubber sheets for balloons, condoms and the like.

The improved rubber was versatile. The emerging electrical industry found it good for insulating wires. Vehicles used rubber shock absorbers. Inflatable rubber tyres (p. 105) found ready markets in bicycles and then motorcars (p. 117). Although Goodyear's invention enriched others, he died a poor man, mostly from costly fights against patent infringements.

Natural rubber is little used nowadays, having been replaced in many applications by twentieth-century synthetic rubber and materials like polythene. But in its day it was a revelation.

In Search of a Safer Match

John Walker, the Lundström Brothers

Traditional ways to get a fire going, always a priority, have included rubbing one piece of wood against another to make heat through friction, and striking a piece of iron on flint to liberate sparks. Nowadays we use matches.

A Frenchman called Chancel developed a match in 1805, with the head a mixture of sulfur, potassium chlorate, sugar and rubber. To light the match, you dipped it into a small bottle of sulfuric acid, not a safe thing to carry. In the 'Prometheus' match, on sale in England from 1830, a tiny glass vial of acid on each match was crushed with pliers to set the match alight; again, not very convenient and, like Chancel's match, it did not catch on.

A more recognisable 'friction match', ignited by striking against a surface, was the brainchild of English chemist John Walker in 1827. Patented by Samuel Jones (Walker did not patent the idea although urged to do so), the matches were called lucifers ('light-bearers'), a term used for many decades even when the technology changed. The smell of burning was unpleasant, like fireworks, and ignition could be violent or even explosive, but the matches were popular and apparently increased the number of tobacco smokers, previously denied a convenient 'light'.

Frenchman Charles Sauria added white phosphorus to the mixture in the head in 1831 to remove the smell. The matches, called congreves after a popular military rocket, and marketed in a sealed box, were cheap and sold well, but workers making them were exposed to phosphorus fumes and developed serious bone disorders, including 'phossy jaw'. Users were at risk too; white phosphorus is poisonous. Ultimately, white phosphorus in matches was banned.

The modern safety match came to light in 1844, later commercialised by the Swedish Lundström Brothers. 'Safety' has two meanings. Much safer red phosphorus replaced the white version, and the various chemicals were distributed between the head of the match, (antimony sulphide and potassium chlorate) and the striking surface (a mix of powdered glass and red phosphorus). Heat from friction in the act of striking turns a small amount of red phosphorus into white. This immediately ignites and sets the head burning.

This is basically the match we use today. Combining the matches and the striking surface into a 'matchbook' dates from the 1890s in the USA.

Now the Blind Can Read

Louis Braille

An accident in his shoemaker father's workshop in 1813 blinded Louis Braille at the age of four. His future was greatly clouded as a result. Unable to read and write, he would be denied an education and most likely reduced to begging to support himself.

Braille escaped that fate by being enrolled in the only school for the blind in Paris. The founder of the school, Valentine Huay, appalled by the treatment of the blind and their lack of opportunities, had devised a way to teach reading using cards on which letters had been strongly embossed and could be felt with the fingers.

The still-young Braille became determined to find a better method, perhaps one that would allow the blind also to write. His immediate inspiration came from an army officer, Charles Barbier, who had visited the school when Braille was only 12. His 'night writing', a system for writing messages that could be read in the dark, did away with the need for a light to read by; soldiers did not need to converse, perhaps giving away their positions. The system used 12 raised dots in a code to represent various sounds.

Barbier's system had proved too complex for soldiers and the army rejected it. However, Braille experimented with it, and devised his now famous system of six dots, punched into paper with a stylus and template and able to convey the letters of the alphabet, numbers, diphthongs and punctuation marks. With training, a blind person could quickly sense the 'letters' as they passed beneath their fingers, restoring not only an ability to read but also to write. Music, mathematics and (nowadays) computer codes are other languages easily translated into braille. The first braille book was published in 1827.

Braille's system was initially opposed by sighted teachers who could not see its benefits. Though later becoming a teacher at the school, he did not live to see braille widely adopted, dying from tuberculosis at the age of 43, poor and largely friendless. Braille writing might have died with him but for the commitment of a small group of blind men in England. In 1868 they founded the British and Foreign Society for the Improving of the Embossed Literature for the Blind, later the Royal Institute for the Blind—now the largest publisher of braille in Europe.

The Railway Revolution

George Stephenson

By the mid-1820s, railways were taking over the life of George Stephenson (p. 83). He had had some successes in other fields, such as his 'safety lamp' able to be used in mines where explosive 'fire damp' (methane) seeped from the coal seams. In his lamp, the flame was shielded by metal gauze that stopped it igniting the surrounding methane. Humphry Davy, the leading English scientist of his day, had the same idea around the same time, with resulting controversy over who had invented what when.

But trains were the thing. Early success in steam locomotive building resulted in a factory to make them. He had built several railways for mines and coal haulage, including a public line between Stockton and Darlington. The engine he built for it could reach the unheard-of speed of 25 kilometres per hour.

The big triumph came in 1829, when he was chief engineer on the proposed railway from the industrial city of Manchester in the English Midlands to the port city of Liverpool. The sponsors of the railway had not been impressed by the performance of the early engines; they preferred horses to haul the carriages. Passengers on the Stockton/Darlington track had travelled in horse-drawn carriages, though goods travelled by steam.

Under pressure from Stephenson,

determined that steam be given a chance to compete, a £500 prize was offered for an engine at least as good as a horse. Stephenson's own entry, the *Rocket,* built with the assistance of his only son, Robert, easily outpaced the opposition in the famous Rainhill trials, covering 100 kilometres at an average of 20 kilometres per hour. Among other innovations, Stephenson used a new style of boiler, with the water sent through the heat of the fire in a number of tubes, so that it boiled faster and delivered higher pressure steam.

After five years of construction, the Duke of Wellington opened the line in 1830. The occasion was marked by tragedy: the local MP, the Honourable William Huskinson, fell in front of the train and became the first person to be killed on the railways.

The age of railways had come; enterprising businessmen were soon assembling capital, buying up land and putting labourers to work. Over the next few decades, railway lines appeared all over England. Travel times tumbled; once the track was laid, Edinburgh could be reached from London in 50 hours rather than 12 days by coach or horse. The nation's leading engineers were engaged. For example, Isambard Brunel, later the eminent builder of the massive steamship *Great Eastern* (p. 99), was chief engineer on the Great Western Railway from London to Bristol.

Stephenson's innovations settled many features of railway technology still seen today. Needing a flange to keep the wheels on the rails, Stephenson had opted for a flanged wheel rather than flanged track; others followed suit. The gap between the tracks ('standard gauge' of four feet, eight and half inches in the old measure) was taken over from the traditional spacing of tracks in coalmines. By now the rails were made of wrought iron (p. 55), less brittle than cast iron rails and allowing faster travel. Stephenson was also quick to promote the use of tunnels, cuttings and bridges to reduce gradients and curves on railways

Railways appeared quickly elsewhere. A steam-powered passenger service was running on the South Carolina Railroad in the USA soon after the Liverpool to Manchester railway opened. Nearly 200 kilometres long, it was judged 'a wonder of the world'. US trains soon developed their distinctive features: the 'cow catcher' to clear obstacles off the track, the bell and heavy whistle, the substantial cabin to protect the crew from severe cold, the sandbox to support traction on steep gradients, and small wheels in front of the main driving wheels to guide the train on a steeply curving track.

Other nations were soon active, though some early trains were horse-drawn. France had its first trains in 1829, Germany in 1835, Russia in 1837 and Italy in 1839. Stephenson himself built railways in Belgium and Spain. An early symbol of the looming dominance of the railroad was the completion of the track across the USA coast-to-coast in 1869. Railways would rule long-distance travel for the most of the next century, though steam would yield to other forms of power.

The triumph of the locomotive drove improvements in materials and manufacturing techniques, and promoted other innovations as well: the 'electric telegraph' (p. 97), first used for railway communications; and the development of 'standard time' (as opposed to 'local time') so that the timetables could be coordinated. There were technology losers too: the extensive network of canals, built over the previous century to carry heavy goods, was now too slow to compete.

Bringing in the Harvest

Cyrus McCormick, John Ridley, Hugh McKay

In 1831, 90 per cent of Americans were 'on the land'. Today, the USA meets its needs for food and fibre through the labours of only 2 per cent of the population. That transformation is due mostly to mechanisation. One machine, conceived in 1831, ultimately increased tenfold the productivity of farmers growing grains like wheat, rye or barley. The slowness of harvesting such crops limited how much a farmer planted. Using a model left incomplete by his father, Virginian Cyrus McCormick took only six weeks to design, build and test a reaping machine. By replacing men on foot using scythes and rakes with a horse-drawn scythe that piled the cut grain onto platforms, McCormick enabled one farmhand to do the work of five.

Local farmers were wary of innovation; some thought the new iron ploughs 'poisoned the soil'. A decade passed before McCormick's willingness to offer credit and his talent for publicity, together with the ready availability of spare parts, and guaranteed productivity ('15 acres a day!') won them over. He sold two machines in 1840, seven in 1842 and 50 in 1844, and had to build a factory in Chicago to meet the demand. In 1851 his machine won a gold medal at the Great London Exhibition.

McCormick continued to innovate. His later models cut, bundled and bound the sheaves of grain, ready for transport to the thresher. They needed just one man to drive the horse (later a tractor). Despite inevitable court challenges to his patents, competition from other manufacturers and the loss of the factory by fire, he won out in the end, founding the International Harvester Corporation.

McCormick's machine cut the crop with the heads of grain intact; threshing took place elsewhere. Another strategy 'stripped' or 'headed' the crop, removing the ears of grain while leaving the stalks standing. Australian innovators John Ridley and Hugh McKay made major contributions here. Ridley's 'stripper', devised in 1843 to counter a labour shortage during a good harvest, held the heads in combs and beat off the grain. He sold tens of thousands of these machines. Forty years later, McKay's 'combine harvester' stripped the heads, then threshed, winnowed and bagged the grain. The Sunshine Harvester factory outside Melbourne became the largest such in the southern hemisphere.

Combine harvesters today are usually self-propelled and may be guided by global positioning systems (p. 277) in large fields, but their basic operating principles remain unchanged.

Electricity on Demand

Michael Faraday, Joseph Henry

Many see Michael Faraday, the bookbinder's apprentice (and protégé of Humphry Davy) who rose to head London's Royal Institution, as the greatest researcher of the nineteenth century, or perhaps ever. His reputation rests mostly on his prodigious studies of electricity and magnetism over a period of 20 years.

He knew of the link the Dane Hans Ørsted had found (p. 88) between electricity and magnetism. An electric current in a wire is surrounded by a magnetic field. In 1831 he asked 'if electricity can produce magnetism, can magnetism produce electricity?' Anyone can do what he did. He wrapped a wire a number of times around one side of an iron ring, and connected the ends to a battery and switch. He wrapped a second coil around the other side of the ring and connected it to an instrument (a 'galvanometer') to report any current that flowed. He hoped to show that a current flowing in one coil would make the current flow in the other, the connection coming via the magnetic field.

Almost nothing happened. When Faraday turned on the current in the first coil, the needle in the galvanometer flicked quickly one way before settling back to zero. When he turned the current off again, the needle flicked the other way and returned to zero. So the magnetic field had to be changing, growing stronger or weaker, before it would make a current flow.

Faraday had found 'electromagnetic induction'. A changing magnetic field 'induces' a current in a nearby wire. The practical consequences were obvious. A machine with magnets and coils of wire constantly moving relative to each other would produce a sustained electric current, as an alternative to using a battery (p. 78). Faraday's first 'dynamo' spun a copper plate between the poles on a powerful magnet. A current flowed in the disc but energy wastage as high. Still the crucial principle had been demonstrated.

A year later French inventor Hippolyte Pixii replaced the disc with some coils of wire, holding them still and spinning the magnet. This began decades of research to find the best arrangement of coils and magnets, including electromagnets, to convert movement into electricity. Some designs produced a current in only one direction (direct current or DC); other designs produced alternating current (AC), which flowed back and forth.

Similar findings were being made across the Atlantic. Natural philosopher Joseph Henry was a major figure in the growth of American science in the nineteenth century, with achievements as both a researcher and a statesman of science. In 1846 he left a post at what was later called

Princeton University to become the first Secretary of the newly founded Smithsonian Institution in Washington, today one of the world's great museums and laboratories in science and technology. He also helped found the National Academy of Sciences, the North American equivalent of the Royal Society.

These roles took him away from his own research but his reputation was already secure, particularly in electricity and magnetism. In the late 1820s he built on the discovery of electromagnetism and made powerful electromagnets by wrapping thousands of loops of current-carrying wire around iron bars. Some of these powered the first electric telegraph in America (p. 104) which, among other benefits, allowed the rapid reporting of weather information and forecasts; Henry was instrumental in setting up a national system for that purpose.

Working independently, Henry built electric motors and discovered electromagnetic induction in the same year as Faraday. He took the matter further, discovering 'self-induction' before Faraday did. Changing current in a coil of wire makes a changing magnetic field; that field in turn generates another current in the same coil opposing the first. So changing currents (including alternating ones) are impeded or 'choked' as they try to pass through a coil of wire. The more rapidly the current changes (the higher its 'frequency') the more noticeable the effect. 'Chokes' or 'inductors' remain vital components of electronic circuits today. Henry's contribution is commemorated in the unit of inductance: the henry.

Faraday and Henry both grasped the use of induction in a 'transformer'. If two coils are wound on the same piece of iron, the first with a few turns of wire, the second with many, the induced current in the second coil has a much higher 'voltage' than the first. So voltages can be 'stepped up' (or 'stepped down' if the coils are reversed) but only with alternating currents (AC); steady 'direct' currents (DC) are not affected. The current in the second coil is much smaller than the first; otherwise energy would be created. Raising the voltage lowers the current, and vice versa, an important consideration in the later 'Battle of the Currents' (p. 151) as the use of electricity began to spread.

Messages over the Wires

William Cooke, Charles Wheatstone

The 'electric telegraph' ('distant writing') was a possibility once Englishman Stephen Gray showed in 1729 that an electric charge could pass down a wire, but a century elapsed before it was a reality. The key problem was how to embed messages in the flow of charge. Early proposals needed 26 wires in parallel, one for each letter of the alphabet. In 1774 French inventor Georges Lessage set up such a system between two rooms of his house. Over larger distances, a practical device would demand many fewer wires.

Users also needed to know if a signal had arrived. Lessage relied on an 'electroscope'; a small pith ball moved as a pulse of charge came down a wire. Once electrolysis was discovered, inventors proposed running the wires into phials of acidified water and watching for gas bubbles released by the inbound current. None of these schemes seemed practical.

Electromagnetism, not known until 1820 (p. 88), was much more promising. The pulse of current would signal its arrival by moving a small compass needle. That idea formed the basis of the first commercial electric telegraph, patented in 1837 by Britons William Cooke, a former army officer, and physicist Charles Wheatstone. The alphabet was arranged above and below a row of five needles;

any combination of two indicated a particular letter. Only five wires were required, and the Great Western Railway Company (GWR) became interested. The burgeoning railway networks needed some communication system to control the movement of trains.

Cooke and Wheatstone's telegraph was installed along 20 kilometres of GWR track west from London in 1839. Some years later, it helped capture John Tawell, who had killed a woman near Slough and fled toward London dressed in the then distinctive garb of a Quaker. This information was telegraphed ahead, but with the description spelt 'Kwaker', since the system did not code for the rare letter Q. Tawell was apprehended following his arrival at Paddington Station, and ultimately hung.

The inventors later fell out over claims by Wheatstone that the telegraph was all his idea. Some dissension might be expected, given the popularity of the telegraph and the revenues beginning to flow. By 1852, 5000 kilometres of telegraph line had been strung alongside the expanding railways and elsewhere. In France and some other countries, the electric telegraph spelt the end of the innovative semaphore systems running since the 1790s (p. 68).

Steam Drives the Hammer

James Nasmyth

Scottish-born James Nasmyth got his start as a protégé of the influential London-based engineer Henry Maudslay (p. 77), as did Joseph Whitworth (p. 103). The association, though impactful, was brief; Maudslay died only two years after he took on the 21-year-old Nasmyth as an apprentice. Nasmyth's father was an eminent landscape painter whose hobby was mechanics. In the elder Nasmyth's workshop, young James took to building models of the newly popular team engines. It was those model engines that impressed Maudslay into giving him a job.

Leaving London after Maudslay's death, Nasmyth set up his own factory, first in Edinburgh, then in Manchester, beside the new Manchester to Liverpool Railway (p. 92). Orders came in briskly, including for machine tools to build the engines of the huge steam ship *Great Britain,* then under construction (p. 99).

Nasmyth's great contribution to heavy engineering was the 'steam hammer', able to shape very large objects from one piece of hot wrought iron, such as ships' anchors or the 1-metre-diameter drive shaft for the proposed paddle-wheels on the *Great Britain* (though the shaft was never needed due to changes in the ship's design). Existing 'tilt hammers', essentially a hammer on a long arm pivoted at the lower end, could not be raised high enough to exert sufficient force to forge such big pieces, which therefore had to be made in small sections and welded together—this was time-consuming and costly.

Nasmyth's insight was to use steam pressure to raise a massive iron block, which could weigh a tonne or more, much higher than previously possible. Steam also boosted the fall of the hammer under gravity to deliver an immense blow to the iron being banged into shape (and to make a thunderous noise!). He also built a pile driver using the same idea.

The steam hammer, devised in 1837, transformed heavy industry: costs were halved, quality improved greatly, previously impossible projects became feasible. Many different products could be formed under the hammer, from iron plates for the sides of warships to parts of steam engines, to large cannons.

Nasmyth was tardy in patenting his steam hammer; he did so only after seeing a replica in operation in a rival's factory in France. Yet he made a good living and was able to retire at 48, devoting himself to astronomy and other interests. 'I now have enough of this world's goods', he said. 'Let the young have their chance.'

Steam Across the Oceans

Samuel Cunard, Isambard Kingdom Brunel

Once steam ships in increasing numbers were carrying cargo and passengers along the inland waterways and sheltered coastlines of Europe, North America and elsewhere (p. 81), the challenge became to send 'steamers' onto the open ocean, such as across the Atlantic. That would require engines using less coal so the ship could stay at sea for several weeks, and replacing paddle-wheels for propulsion with something less affected by the rolling of the ship.

Without waiting for such break-throughs, crossings under a combination of steam and sail got underway in 1819 with the American ship *Savannah*. A regular service took another two decades and introduced the famous name of Cunard. Securing the British government contract for the mail service across the Atlantic, Samuel Cunard established a shipping line in 1840, soon carrying passengers as well and offering guaranteed sailing dates. Cunard's first ships used the sail/steam combination but the era of the passenger liner, using steam alone, was drawing to a close.

When it came to building the ships, Isambard Kingdom Brunel set the pace. The brilliant and daring Brunel had already built the Great Western Railway. He created ever-bigger, faster ships that were more comfortable for passengers. The *Great Western,* launched in 1838, was 70 metres long and crossed from Bristol to New York in just 15 days.

His 1853 *Great Britain,* nearly 100 metres long and luxuriously appointed, was the first ocean-going steamship made of iron, and the first to use the submerged 'screw propeller' for propulsion in place of paddle-wheels. The idea of the screw had been around since the experiments of the American John Stevens in 1803 but only in 1838 had a large steamer used one—the riverboat *Archimedes*, built by screw pioneer Francis Pettit Smith. Later ships had twin screws for reliability.

Brunel's 1858 *Great Eastern* was too large, at 200 metres plus and 19 000 tonnes, to succeed commercially, though it did successfully lay the first transatlantic cable (p. 115). But it led the way in technology, as shipbuilding moved from iron to the more flexible steel for construction, and to 'double-expansion' engines. Steam exhausted from one cylinder drove a second larger one, so that discarded steam contained less energy. This cut coal usage. Crossing times were steadily cut, to below 10 days and then below six, as shipping lines sought the coveted 'Blue Ribband' for fastest passage.

Paying
for the Post
Rowland Hill

Innovations go well beyond machines. They include systems and procedures that satisfy vital needs. The postal service is an example; the experience in Britain was typical. The Master of the Post, appointed in 1516, dealt only with official correspondence. In 1635, an Inland Postal Service began between London, Edinburgh and the important seaport Plymouth, and King Charles declared the 'Royal Mail' now open for public use.

Within a few decades, postal routes using 'stagecoaches' linked the cities and big towns. Based on both the distance carried and the weight of the letter, charges were high; only businesses and the wealthy used the mail service. In 1680 'penny post' was introduced for letters delivered around London, paid in advance by the sender; previously the receiver bore the cost. All revenue from the steadily expanding service went to the government. To help pay for the Napoleonic War, the cost of London 'penny post' was doubled in 1801.

The arrival of railways (p. 92) speeded the mail, but the most significant innovation came in 1840. Former schoolmaster and colonial administrator Rowland Hill convinced the Government to introduce 'labels', small adhesive squares stamped to record the postage paid. Soon called 'stamps', these were initially cut from a page with scissors. Perforated stamps, able to be easily torn off the sheet, came in 1854. Already the authorities were busy devising ways to prevent stamps being used more than once.

The first stamp issued was the 'Penny Black', prized by stamp collectors today as it did not remain long in circulation. That sent a small letter anywhere in the UK regardless of distance. By his reforms, Hill wanted to increase efficiency and keep down costs so that everyone could afford to use the mail. Usage doubled, but the uniform rates slashed revenue, which took 35 years to recover.

Other nations quickly adopted the practice: Brazil in 1843, Switzerland in 1845, the USA in 1847, France in 1849. The rapid globalisation of mail services required some international collaboration, and the international General Postal Union was established in 1874.

Though letter mail now competes with facsimile and email (p. 270), the massive growth in 'snail mail' over recent decades has seen hand sorting replaced by mechanisation. Modern letter-sorting machines rely on 'post codes' or 'zip codes', in place across the planet. Hill had made the first such steps in 1847, dividing London into postal districts by compass direction (NW1, SW3 and so on).

More Rounds a Minute

Nicholas von Dreyse, Samuel Colt, Richard Gatling, Hiram Maxim

Gun technology had been slow to change: in 1840, 300 years after their invention, single-shot flintlock muskets, rifles and pistols (p. 13) were still in use. Then quite quickly, new technology arrived. By 1900 guns used 'cartridges', combining charge and bullet and ignited by percussion, not by a spark. Bullet and charge were loaded from the rear ('breech-loading') rather than down the barrel ('muzzle-loading'). Guns could fire several shots, sometimes many, in rapid succession.

The replacement for the flintlock was the newly discovered mercury fulminate, which ignited with force when struck. A small canister of this chemical replaced gunpowder in the flash-pan. Pulling the gun's trigger made a 'hammer' strike the canister; the resulting small explosion ignited the main charge in the barrel. This invention had a number of claimants, including Scottish clergyman John Forsyth and American landscape painter Joshua Shaw; the first patents were secured late in the 1820s.

Cartridges and Bolt Action

Soon after came the 'cartridge', a small paper package containing the bullet, the charge of powder and the fulminate detonator, inserted into the gun barrel together. The famous 'needle gun' with its 'bolt action' was invented by the Prussian gunsmith Nicholas von Dreyse. The prone rifleman pulled back the bolt, opening a small slot in the barrel, and inserted a cartridge. Advancing the bolt sealed the barrel again; pulling the trigger drove a long needle into the detonator and the gun fired. These breech-loading guns could fire 10 or more rounds per minute, compare to only three with muzzle-loading guns.

Breech-loading guns became standard in the Prussian army from 1840. The wide-barrelled Dreyse gun had low muzzle velocity, and was out-ranged by better quality muzzleloaders, but the rapidity of fire proved decisive. Within decades, all armies were equipping with bolt-action rifles, standard infantry issue for the century to come, with only the addition of 'magazines' holding a small number of cartridges. These did not have to be loaded one by one, so fire rates increased still further.

There were major consequences. Battles became bloodier. Daylight cavalry charges and infantry manoeuvres ceased. No longer needing to stand to load, riflemen could lie hidden from sight. Increasingly, victory on the fields went to the side using the new technology; this hastened the rise of Prussia as a military power. Mass production made guns faster and cheaper to manufacture, and so more widely available. Away from the battlefield, the harvesting of wild animals, such as American buffalo and African elephants, became more brutally efficient.

Revolvers

Repeating guns were already on the market, the most famous being American Samuel Colt's 'revolver' patented in 1836. Colt struggled at school, becoming a sailor when only 13 and later a travelling showman dispensing 'laughing gas' (p. 107) to volunteers for the amusement of all. He had devised his repeating gun at sea and built a wooden model.

The revolver was slow to gain popularity. Colt's company making them was bankrupted in 1842, but the outbreak of war with Mexico in 1846 suddenly increased demand. Colt learnt the lessons of mass production from Eli Whitney (p. 71) and was soon making Colt revolvers by the thousand, and a fortune for himself, in the world's biggest arms factory on the banks of the Connecticut River.

Colt's revolver was ingenious. Holes drilled in a barrel-shaped chamber held five or six shots, initially percussion caps and balls, later complete copper-clad cartridges. Pulling the trigger fired one round; releasing it rotated the chamber so that another shot was brought into the barrel. Shots could now be fired off almost as fast as the trigger could be pulled. Although its use in the 'wild west' would

later be overstated by Hollywood, the Colt 45 was a powerful influence in both maintaining and undermining law and order.

Machine Guns

The next step would dispense with regular pulls on the trigger. A true 'machine gun' would continue to fire until the trigger was released. The 'Gatling gun' invented by American Richard Gatling in 1861, in time for the Civil War where it increased carnage on the battlefield, fell short of that ideal. It had 10 barrels, rotated by a hand crank, with bullets fed in by gravity from a drum. It jammed rather too easily but could fire 300 rounds a minute.

Gatling's fellow American, Hiram Maxim, found the answer in 1881. Recalling the 'kickback' of a rifle from his youth, he devised a single-barrelled gun using that recoil energy to reload the gun after each shot, the bullets being fed in on a belt. Though innovative, the gun would not have worked without cordite, the more powerful explosive invented by Englishman James Dewar (p. 150), which gave sufficient kick to drive the mechanism. The 'Maxim Gun', running so hot it needed water-cooling, first saw serious service in the Russo–Japanese war of 1904, but its real harvest would be secured in the trenches of World War I.

The Standard Thread

1841

Joseph Whitworth

The legacy of British engineer Henry Maudslay, inventor of the modern metalworking lathe (p. 77), was powerful, and included his many apprentices and co-workers, who went on to illustrious careers and important inventions of their own. One such was Joseph Whitworth, a name remembered today.

Whitworth was mechanically minded from an early age and served as an apprentice in several machine tool companies, including that of Maudslay in London. He opened his own workshop in Manchester in 1833, making lathes and other machine tools renowned for high-class workmanship.

His great and memorable achievement lay in standardising the sizes of threads cut into the screws and bolts that held machines, ships and the like together. This may sound a routine, even dull, matter but its impact on industry, and therefore on production and living standards, was immense. Until that time, anyone cutting a screw thread on a metal rod could make it how they chose, any number of threads to the inch, any angle of the walls of the thread. Mass production was impossible. Nuts and bolts were generally not interchangeable, having been made as unique pairs. Lose the nut and you might not find another to fit the bolt.

Whitworth proposed that every screw of a given diameter should have the same number of threads to the inch (eight per inch in the case of a 1-inch screw) and that the angle in the thread should always be 55 degrees; he set the latter figure by sampling screws from a number of leading manufacturers and taking the average. Whitworth submitted his proposal to the Institution of Civil Engineers in 1841 and this scheme was adopted throughout Britain in 1860. His name, as in British Standard Whitworth (BSW) lives on as a result, or did until the UK went metric.

Whitworth also pushed the technology of measurement to unheard-of precision. Building on the lessons from Henry Maudslay's 'Lord Chancellor' device, Whitworth's micrometer could detect differences in length of one-millionth of an inch. This was such a wonder that it went on display at the Great Exhibition in London in 1851.

Unlike some, Whitworth did very well from his engineering business, in terms of both recognition (he was, for example, a Fellow of the Royal Society) and financial reward (he owned a large country house in Derbyshire). The proceeds of his estate continue to support philanthropic works.

The Telegraph Expands

Samuel Morse

American painter and inventor Samuel Morse is often named 'creator of the electric telegraph'. He was not; others had been busy before him (p. 97). He did devise a better way to send and receive messages. Early versions required multiple wires; even Cooke and Wheatstone's successful equipment needed five. Morse knew just two wires would suffice if letters and punctuation were represented by a code. The most common letters, such as 'E', would use the simplest combination of symbols.

Morse and his backer Alfred Vail demonstrated their creation, with its combinations of short 'dots' and long 'dashes', before a scientific committee in 1839. 'Morse code' was an early example of 'binary code', now universal in information technology. The first telegraph line in the USA, government funded, connected Baltimore and Washington in 1844. Morse himself keyed in the first message, the biblical quotation 'What hath God wrought?'

It was suitably portentous. As in Europe, telegraph wires spread rapidly across the continent. By the 1860s a telegraph message could go coast to coast, ending the brief era of the 'pony express'. That expansion was paralleled by the growth of lawsuits as Morse fought to protect his patents. The various companies he and others established ultimately merged to form Western Union.

The first Morse/Vail system marked the dots and dashes on a moving paper tape. Operators soon realised they could read the message from the clicking of the electromagnetic equipment; the tape was unnecessary. Morse devised 'relays', that effectively amplified the signal and sent it over ever-increasing distances.

As the decades passed, morse code was supplemented by one using dots and spaces in groups of five. Such 'letters' could be punched into paper tape, automating transmission and allowing for messages to be prepared ahead of time. Still other technologies, devised by Thomas Edison (p. 135) among others, allowed 'multiplexing', sending more than one message at a time. The productivity and convenience of the telegraph multiplied, costs plummeted, messages sent 'by wire' became commonplace.

Though revised in 1848 to become 'International Morse', morse code is mostly now obsolete, a historical curiosity, like the electric telegraph itself. Yet it began the shrinking of the world into a 'global village', where information and personal contact is only a few seconds away. At the end of the century, wires were no longer essential; the 'wireless telegraph' was the first step towards today's radio (p. 160).

Riding
on Rubber

Robert Thompson, John Boyd Dunlop

Today we travel smoothly by car, bus, motorbike or bicycle thanks to the pneumatic (air-filled) rubber tyres on the wheels. The name associated with this source of comfort is usually John Dunlop; certainly Dunlop tyres are still sold around the world. Like many such linkages, it is only partly true.

Scottish-born John Boyd Dunlop was a successful veterinarian in Belfast. One day in 1888, his young son wanted something to put around the solid wheels of his tricycle to smooth the ride over cobbled streets. Dunlop had some skill working with rubber, and made ring-shaped rubber tubes that fitted over the wheels and were filled with air.

The benefits were obvious. Dunlop sold some of these new tyres to local cyclists, who won more races using them (less friction). The head of the Irish Cyclists Association partnered Dunlop in marketing the tyres. The company prospered, but Dunlop sold out too early and didn't leave much in his will.

All true, but incomplete. Unknown to Dunlop, the pneumatic tyre already existed, invented by fellow Scotsman Robert Thompson 40 years earlier. The idea was basically the same; an annular tube of rubberised canvas, filled with air and protected by an outer covering of leather. Bicycles hardly existed then (p. 117) and Thompson's tyres supported horse-drawn wagons.

The rubber industry was new— vulcanised rubber was barely discovered (p. 89)—and Thompson could not secure enough rubber of the right quality and thickness. So he turned to solid rubber tyres, initially for wheelchairs. Four decades later, Dunlop had the right rubber and his inflatable tyres reached the market, but he could not gain a patent because Thompson had precedence.

Unlike one-idea Dunlop, the inventive Thompson's portfolio included: a fountain pen; a steam-powered road tractor (with solid rubber tires) able to pull a 40-tonne load and later to make the draught horse obsolete on farms; and an improved 'mangle' for squeezing the water from clothes, able to turn either way and invented when he was only 17. His technique for setting off explosive charges safely by an electric current saved lives in mining and construction, his improvements in sugar refining (undertaken while on assignment in Java) greatly enhanced productivity. Thompson, who had worked with railway genius Robert Stephenson (p. 92), was inventing up to his death in 1873 at only 50. His last patent (posthumous) was for an inflatable rubber cushion.

Restoring the Smile

False Teeth and Toothbrushes

Replacing teeth lost by decay or accident, or simply enhancing the appearance of existing teeth, is an advanced technology and a highly profitable industry today, but it has a long history. Two and a half thousand years ago, the Etruscans were skilled with both 'crowns', replacing a single tooth, and 'bridges' to span several missing teeth. Such technology was later forgotten.

For centuries rampant decay led to much tooth loss. England's Good Queen Bess carried a piece of white cloth in her aging mouth to camouflage her missing front teeth. False teeth were consequently much in use, carved from ivory or bone, embellished with gold and occasionally set with real teeth extracted from the poor for payment of a fee, or from cadavers, including soldiers killed in battle. Attached by silk threads to existing teeth, these were purely ornamental. Wearers had to take them out to eat, they never fitted well, and they were often turned rancid and brown by mouth juices.

Teeth you could seriously chew with came in the mid-nineteenth century. Several technologies came together: the teeth themselves, cast individually in porcelain (p. 39) by fastidious dentists who detested handling 'dead men's teeth'; flexible plates made from the new

'vulcanised' rubber (p. 89) acting like gums to hold the teeth and accurately cast to fit the mouth accurately; and the use of nitrous oxide (p. 107) to dull the pain of tooth extraction. More people had their decaying teeth extracted, increasing demand for 'dentures'. With the better fit, even full upper plates would stay in place; earlier upper sets had to be held up by springs, which sometimes caused them to bound right out of the mouth unexpectedly.

The need for false teeth could be reduced by taking better care of real ones. Ancient cultures used 'cleaning sticks', softened by chewing. Bristle toothbrushes, probably invented in China, reached Europe in the seventeenth century and were much advocated by French dentists, who led the rest. Nylon replaced natural bristles in the mid-twentieth century; electric toothbrushes came along around 1960.

Toothpaste replaced older abrasive tooth powders early in the eighteenth century; chalk 'crushed limestone' became the chief ingredient mid-century (and remained so); the collapsible tube, lead initially, aluminium later, came about the same time. Flavourings and additives for various purposes led to a wide diversity of products; fluoride was first marketed in 1956 to Proctor and Gamble's 'Crest' toothpaste.

Anaesthetics Against Pain

William Morton

Anaesthesia, the controlled use of chemical compounds to reduce pain, was one of two nineteenth-century breakthroughs that made modern surgery possible, the other being control of infection.

Previously, pain could be relieved by alcohol (as in rum), opium (laudanum), mandrake, cocaine, perhaps a blow to the head to render the patient unconscious, pressure on nerves, intense cold; none were very effective. Even into the nineteenth century, most operations took place without any anaesthetic, the surgeon relying on speed and skill to reduce the trauma of pain to the patient. There was little time for care or complexity.

The first chemical found to reduce pain, nitrous oxide, was initially used as entertainment at parties, hence its common name 'laughing gas'. English chemist Humphry Davy had discovered its exhilarating and relaxing effects in 1800, and predicted it would be used in minor operations, such as the removal of teeth. But that did not happen until 1844, when it was first employed by Horace Wells, a Connecticut dentist.

Wells' work inspired a former student, William Morton. He found that sulfuric ether, known since the sixteenth century, deadened pain when inhaled; it also put patients to sleep. After animal trials, he first anaesthatised a human patient in 1846, during an operation on a tumour on his jaw. Others had been experimenting with ether; Crawford Long of Georgia used it on a patient in 1842, but kept his work quiet until Morton's became known.

Anaesthesia crossed the Atlantic late in 1846; Robert Liston, trained in the USA but working in London, used ether during an amputation. The news encouraged others, including Scottish midwifery specialist James Simpson. He preferred chloroform, more pleasant for the patient and able to be administered by a few drops on a gauze over the nose. He found it useful both for surgery and obstetrics, the practice gaining popularity after Queen Victoria allowed it during one of her confinements.

Since that time, anaesthesia has seen much technological improvement: new anaesthetics, new techniques to administer them, injectable painkillers, epidurals and the use of curare (since 1942) as a muscle relaxant. Still, it seems right to cite 1846 as the key date and William Morton as the pivotal central figure. As his headstone records:

Inventor and revealer of inhalation anaesthesia; before whom, and in all-time, surgery was agony; by whom, pain in surgery was averted and annulled; since whom, science has control of pain.

Machines that Sew

Elias Howe, Isaac Singer

Sewing is an ancient craft; eyed needles made of bone or ivory were used at least 40 000 years ago. Mechanised sewing might have come sooner than it did if the various independent inventors had come together. As was not uncommon, they struggled for patent rights and recognition. The French tailor Bartholemy Thimonier, who designed the first tolerably efficient machine in 1830, died in poverty. His clumsy wooden machines worked but failed to sell. Some were destroyed by rioters.

The first patent on a sewing machine was granted to 24-year-old American mechanic Elias Howe in 1846. His machine replicated the process of hand sewing with needle and thread, using a lockstitch. It could sew only straight seams, though with unprecedented speed, reliability and evenness.

Howe had hoped to help his young wife with her dressmaking, but instead outraged the tailors of Boston. Fearing for their livelihood, they wrecked the machine he demonstrated to them, and Howe had to flee. He relaunched his invention in Britain. There it met no such opposition and swiftly gained popular acclaim, in part through competitions that matched one of his machines against teams of dressmakers.

Howe soon had rivals back home, notably Isaac Singer, whose name remains associated with sewing machines to this day. Singer added a foot treadle to provide power and used a straight needle able to sew curved steams. The pair was soon in conflict over patent rights. Howe won the court battle, but it made little difference. The demand for sewing machines was so great there was room for more than one supplier.

Singer was a better businessman than Howe, selling his machines on an installment plan to spread the cost for his customers. His company made 43 000 machines in 1867 and an astounding half million in 1880. The sewing machine was the first 'consumer appliance', a machine bought specifically for use in the home, and soon most middle-class homes had one.

The popularity of the machines affected fashion; around the 1870s women's clothing became more elaborate, yet able to be machine-made at home with the aid of paper patterns printed in women's magazines. It affected other industries as well, driving advances in mass production much as gun manufacture was doing. The need for reliability raised the demand for advanced materials with high resistance to wear and tear, such as powder-coated metals.

Machines that Wash

James King, Thomas Bradford

The nineteenth century saw the first serious attempt to mechanise clothes washing, cleanliness becoming more earnestly sought than in previous centuries. Earlier efforts included a wooden cage to hold the clothes and a handle for turning it over and over through the washing water, unveiled in 1782.

In 1851 American James King patented the first machine that looked vaguely modern, with a hand-turned drum. A decade later, Thomas Bradford in Britain had washing machines on the market, with a wringer (a pair of rollers to squeeze out much of the water) above an eight-sided wooden box holding clothes and water and turned by hand.

Nothing much changed until the early twentieth century, when small electric motors (p. 88) began to replace hand power for the tub and wringer; these were initially unsafe as water would splash into the motor and give the operator an electric shock. We also saw the introduction of the central 'agitator', which pushed the clothes through the water while the tub as a whole stood still. Metals began to replace wood. Early manufacturers included still well-known names such as Whirlpool and Maytag, but it was half a century before washing machines were a common household item.

Three major steps transformed those early machines into the washing machines now found in almost every home, or in the nearby laundromat. The first freed the operator from the burden of constant supervision. A mechanical timer started and stopped the machine as required, as in the machine introduced by the Bendix Corporation in 1937. Early automatic washers cost as much as a small car: mass production has now reduced that cost by 95 per cent. Washing machines made since the 1980s have microprocessor (p. 267) control of operations, even the addition of detergents, running through the steps from pre-wash to wash to rinse to spin. It is 'set and forget'.

The reference to 'spin' highlights the second development, the integration of the washing machine with the 'spin dryer'. The latter was invented in the 1880s (p. 136), about the same time as the cream separator, which works on the same principle, and reached the USA around the 1920s. It was forgotten and then reintroduced in the 1960s. Nowadays, we expect any decent washer to spin clothes damp-dry as well.

The third innovation, though not universal, has turned the washer on its side, becoming 'front-loading' rather than 'top-loading', and needing less water for the wash.

Getting Under the Skin

Alexander Wood, Charles Pravaz, William Halsted

'Hypodermic' means 'under the skin'. A syringe with that name has a thin, hollow steel needle, able to inject medications into tissues or blood vessels, or to withdraw blood if needed. Its invention appears to have been shared by Scottish doctor Alexander Wood and French physician Charles Pravaz in the 1850s, working quite independently. Prior to this, syringes such as those used by Edward Jenner in vaccinating against smallpox (p. 72) were crude and clumsy. Among other applications, Wood used the new device to inject painkillers such as morphine (derived from opium) into the tissues of the face to ease the pain of neuralgia.

It did not take long for a downside of the use of hypodermics to emerge. Overuse of easy-to-administer morphine injections to control the pain of battle wounds during the American Civil War in the early 1860s resulted in many soldiers returning home addicted. It is reported that Wood's own wife died from an overdose of self-injected morphine, the first recorded such fatality.

The convenience of the hypodermic syringe encouraged the search for other injectable painkillers. Around 1884, leading American surgeon William Halsted (later the first to use rubber surgical gloves and to perform radical mastectomies as a treatment for breast cancer) promoted the use of injected cocaine (distilled from the leaves of the coca plant) as a local anaesthetic. Experimenting on himself, he became addicted, as did other colleagues. By altering slightly the molecular structure of the drug, chemists later removed its addictiveness, producing novocaine, later widely used in dentistry.

Disposable (use-once-and-discard) syringes of glass were first mass-produced in the 1950s to support the large-scale vaccination of children against polio with the newly developed Salk vaccine (prior to the introduction of the Sabin vaccine taken by mouth). Plastic disposables came soon after. Even more modern syringes are disabled after one use, preventing their re-use. This has helped stem the spread of diseases such as AIDS, associated with injection of illegal drugs using contaminated needles.

The Rise of the Skyscraper

Elisha Otis

Big cities today are dominated by tall buildings. Thirty or 40 stories are not uncommon; real skyscrapers top 100 floors. About 125 years ago, the typical cityscape was very different; few buildings reached more than three or four stories, limited by building techniques, access and the provision of services.

Even then, numerous innovations were encouraging the upward urge. Metal-frame construction, made possible by the new Bessemer steel (p. 116) and enhanced by Joseph Monier's reinforced concrete (p. 125), meant that outer walls no longer carried the weight of a building. Water could be sent to the upper floors with powerful pumps driven by electric motors (p. 88). Communication from the lowest floors to the highest was made easier by the telephone (p. 133). Most significantly, the endless climbing of stairs was banished by Elisha Otis's elevator.

Platforms pulled up and down on ropes had been carrying goods from floor to floor in warehouses and the like for decades, but putting passengers in such lifts raised safety concerns. What if the rope broke? Otis, employed as a youth in a bed-manufacturers warehouse, realised that a fail-safe lift needed some automatic, perhaps spring-loaded, device that could snap into place if needed and lock the platform in the shaft.

In 1853 he demonstrated his confidence in his invention in a striking manner at the New York Crystal Palace Exhibition. Standing several stories up on an open platform, he ordered his assistant to cut the rope supporting it (in some accounts he slashed the rope himself with a sabre). The watching crowd gasped, but Otis did not fall to his death. He stood on the unmoving platform, waving his top hat. It was a marketing triumph.

His 'safety elevator' had lived up to its name. Soon everyone wanted one. Otis went into the lift-making business. By 1873 more than 2000 Otis elevators were in place in office buildings, hotels, apartment blocks and department stores across the USA. In 1904 the Otis organisation, run by his sons (Otis himself had died in relative obscurity in 1861), pioneered the high-speed elevator that real skyscrapers needed.

The term 'skyscraper' had entered the language in the 1880s, when the tallest buildings were merely 20 stories; striking for the day, trivial now. The purpose in going up was, of course, to make more efficient use of precious land space in cities, and to increase its value. The look of our cities changed forever.

Potato Crisps—
The World's Favourite Snack?

George Crum, Herman Lay

When it comes to popular snacks, the potato chip or 'crisp' must be up there with the leaders. Potatoes have been a staple in many European countries since they were brought from the New World centuries ago, and are served in an immense range of styles. Thomas Jefferson, later president of the USA, came to enjoy the French style while on duty as ambassador there in the late eighteenth century. He brought the recipe home and served the thick-cut fried potato slices (not today's meaning of 'French fries') to his guests, many of whom had previously thought potatoes suitable for feeding only to pigs.

In 1853 Native American chef George Crum had French-fried potatoes on the menu at the Sun Moon restaurant at the expensive Saratoga Springs resort in upstate New York. But one diner, reputedly the millionaire banker and New York social icon Cornelius Vanderbilt, did not care for them. Too thick, he declared, and sent them back. Crum prepared a second, thinner serving but the guest was still not satisfied. Irate, Crum cut them so fine they went crisp when fried, too thin and hard to be speared with a fork. He expected the diners to be angry, but they were delighted. Proclaiming the browned, paper-thin tidbits delicious, they demanded more.

Crum was onto a winner. He made his invention a specialty of the house, calling them 'potato crunches'. He was soon packaging them for sale as Saratoga Chips. He ultimately had his own restaurant nearby; Vanderbilt was one of his backers.

Some distance remained to be covered before this New England dinnertime delicacy could be counted a global gastronomic phenomenon. Three key events all occurred in the 1920s: the invention of machines to peel and slice the potatoes, previously done laboriously by hand; the first use of waxed paper bags to keep the crisps crisp (plastic film not yet being available); and the intervention of Herman Lay. A travelling salesman from the north, he peddled potato chips to storekeepers throughout the American south from the boot of his car. He built a business that linked his name indelibly with the salty snacks, especially once he merged his company with Frito, a Dallas-based firm that made corn chips. Frito-Lay is the largest maker of potato chips in the USA and therefore (presumably) on the planet.

Crude Oil Challenges Coal

1854

Benjamin Silliman, Edwin Drake

In the mid-nineteenth century, coal was king, burnt in vast amounts to make steam to drive factories and locomotives. Coke made from coal released iron from its ores in blast furnaces (p. 54). Coal gas lit homes, shops, factories and streets (p. 69). Coal tar (p. 114) was beginning to yield its multitude of chemicals to make dyes, drugs and synthetic materials.

But a challenger was emerging, in the form of petroleum, the flammable 'oil from rock', collected for millennia as it seeped from the ground. Researchers had begun to wonder what petroleum really was. American geologist and opponent of slavery Benjamin Silliman was the first to supply an answer in 1854. Encouraged by New York lawyer George Bissell, who suspected petroleum might contain something useful for lighting, Silliman 'distilled' petroleum from a seep in Pennsylvania. Heating it first gently, and then more strongly, he drove off a succession of vapours with ever higher boiling points, condensing to form liquids.

Petroleum proved to be a mixture of such liquids, from very light ones quick to evaporate, through to some so dense and tarry they were almost solid. The lighter fractions had no immediate use; a slightly heavier one dubbed 'paraffin' or 'kerosene' was a good substitute for the expensive whale and seal oil used in lamps. It was also a solvent for greasy substances that would not dissolve in water. Some of the heavier liquids were good lubricants, able to keep machinery running smoothly and iron free from rust.

Collectively the products made by distilling crude oil were worth having. Holes were soon being drilled into the ground in search of deposits. The first in North America (and second in the world), pioneered by Edwin Drake (and dubbed by skeptics 'Drake's Folly'), hit oil only 20 metres down in Pennsylvania. Soon dozens of drill rigs and pumps pocked the landscape in promising areas; pipe lines and special ships were built to transport the 'crude oil'; and 'refineries' were established to process it into useful products.

The age of oil made a quiet beginning. In time, a vital use would be found for the mix of more volatile fractions that were usually thrown away or remained as a contaminant in kerosene, causing lamps to explode and houses to burn. Renamed 'gasoline' in America and 'petrol' elsewhere, this liquid would set the world on wheels in the century to come.

Perkin's Purple

August Wilhelm von Hofmann, William Perkin

Coal tar does not look very promising: it's a thick, black, smelly, oily mess left behind when coal is strongly heated ('destructively distilled') in a closed container to make coke (for smelting iron ore, p. 54) and coal gas (for lighting and heating, p. 69) Chemists soon guessed that coal tar is not a single substance but a mixture of compounds, some of which could be liberated by gentler ('non-destructive') distillation. If the heating was done carefully and the right temperatures chosen, the various compounds could be driven off and condensed one by one.

An early discovery was the compound aniline and its close relative phenol. The German Friedlieb Runge found those in 1834. Phenol would later be called carbolic acid and valued for its ability to kill the microorganisms that could spread disease. It played a major role in the early growth of antiseptic surgery (p. 126).

Guessing a bit how the colourless, oily aniline was put together, English chemists thought it might be similar to quinine, the only known defence against malaria, then a scourge of the British Empire. In 1845 the leading German chemist August Wilhelm von Hofmann was imported to run the newly established Royal School of Chemistry in London, with the major task of somehow converting aniline into quinine.

In 1856 William Perkin, a laboratory assistant still in his teens, reacted aniline with potassium dichromate and washed the result in alcohol. He found not quinine, but a beautiful purple dye, which he called Perkin's mauve or mauveine. It was the first synthetic dye (other than the excellent aniline black that Runge had patented in 1834), and the source of Perkin's personal fortune. He dropped out of the college to set up a dye-making business with his father's money. Mauveine became immensely fashionable and popular for women's clothing, especially once the empress of France began to wear it. Perkin prospered. Biology benefited too. Perkin's mauve proved very useful in staining specimens and revealing hidden details.

The synthetic or coal-tar dye industry was born. It grew rapidly, especially in Germany once Hofmann returned to his homeland. Synthetic dyes in many other colours were soon developed and, together with medical drugs and other useful products made from coal tar, would be the wellspring of German industrial and economic strength in the twentieth century. But there was a downside: many compounds distilled from coal tar would prove to be cancer causing.

The Telegraph Goes Global

William Thomson (Lord Kelvin)

With swarms of telegraph wires (p. 104) beginning to draw together cities and towns throughout countries and continents, the next challenge was linking the continents by well-armoured and waterproofed cables laid on the seabeds. Britain and France were so joined in 1851 by a cable between Dover and Calais; Britain became a part of the information network spreading across Europe.

As always, the Atlantic was the great barrier. A cable from Ireland to Newfoundland, laid in 1858 by Brunel's massive ship *Great Eastern* (p. 99) soon failed from corrosion and penetration of sea water. A second endeavour in 1866 proved successful, though it required a lot of new understanding of electric currents in wires, and new technology able to make sense of the very faint signals arriving at the far end. Much of this was provided by the great British physicist William Thomson, also known as Lord Kelvin.

By the turn of the century a total of 14 undersea cables sewed the hemispheres together. Messages in morse code could pass between major cities across the planet in minutes or hours rather than weeks and months. A single wire strung across Australia in 1876 connected the cities of the south and east to Darwin and thence to Europe by cables through southern Asia and the Middle East. Messages from the 'Old Country' that had previously taken half a year could now be received as fast as they could be written down. The telegraph became vital in uniting the scattered empires being accumulated by European nations.

Initially the main users of the telegraph (and the new cables) were government departments, police, railways, news reporters, businesses and currency and share traders. In time, ordinary peoople began to take advantage of the new wonder, and to marvel at how quickly news could now spread. With the invention of the telegraphic money order you could 'wire' money to somebody, across the world.

The advent of the telegraph had many spin-offs: for example, it made possible the drawing of synoptic charts, the 'weather maps' on which meteorologists base their prognostications. To draw such a chart, a large number of measurements, particularly of air pressure, had to be taken over a wide areas at the same time and communicated to a central point within a few hours, otherwise the chart would be out of date and useless. The telegraph made that possible.

Steel Replaces Iron

Henry Bessemer

Steelmaking is ancient, dating back many centuries for high-quality goods like swords. Production was labour intensive; the 'Hundred Refinings' method used by the Chinese indicates its complexity and consequent cost. Despite its advantages, steel could replace iron more widely only when it could be cheaply mass produced.

That transformation was mostly the work of Englishman Henry Bessemer, though he drew ideas from American William Kelly; there was argument later over patents. Bessemer conquered the troublesome brittleness of iron by reducing its carbon content. To achieve this, he had air blown through molten iron contained in a large pear-shaped vessel he called a 'converter'. This caused much of the carbon to burn away, releasing heat to keep the iron molten and speed the refining. When Bessemer described the process before the British Association in 1856, iron experts remarked they had never heard of anything 'so striking and impressive'. It still took four years of effort, and some failures, to make the process acceptably reliable in producing steel. He opened his first plant in Sheffield in 1860.

Bessemer steel soon overtook wrought iron, used for many decades since its invention by Henry Cort (p. 55). Steel matched wrought iron for most purposes and could be made in larger quantities for half the cost. In 1850, 50 000 tonnes of steel were made in Britain by a slow and costly process. Thirty years later, production of steel was over a million tonnes, with Bessemer converters making 80 per cent of that.

Military needs drove demand. The Crimean War of the 1860s showed that steel outperformed iron in cannons and armour-plating; steel for guns had been Bessemer's initial goal. Steel was attractive also to builders of railways (steel rails were more durable than wrought iron ones) and ocean-going ships. Steel girders soon supported buildings, including the new 'skyscrapers' (p. 111).

Famous for his converter, Bessemer was also active in the sugar industry, steamboats and big telescopes. Honours included a knighthood and the Fellowship of the Royal Society. He earned £5 million for his steel process alone. (He already had a fortune. When still young, he had powdered brass and sold it as an additive to make 'gold paint, much in demand for gilded decorations.)

In time, Bessemer's method was inevitably superseded, initially by the 'open hearth' process developed in Germany, though Bessemer converters were running in Britain until the 1970s. Bessemer had enabled steel to replace iron and that was what mattered.

Bicycles at Last

Pierre Michaux

Why did the invention of the bicycle take so long? For 5000 years, wheels had been used in various combinations, including two side by side. But no one had put two wheels in the one line and sat between them until the Frenchman de Sivrac around 1690. Yet it did not take on. For one thing, de Sivrac could not steer his machine.

Around 1816 the German inventor Karl Drais marketed his 'running machine' or 'Draisine', often dubbed a 'hobbyhorse'. Steerable, it had no pedals, the rider propelling the machine by kicking the ground on either side. It was not a comfortable ride; the wooden wheels had iron rims. Yet, satisfied with his design, Drais made no efforts at improvement.

In 1839 the Scotsman Kirkpatrick Macmillan connected the back wheel to pedals with cranks, and a rider could easily move faster than they could walk. But the bicycle boom did not come until the French carriage-maker Pierre Michaux, asked to repair a Draisine, proposed a pair of pedals fixed to the front wheel, producing the 'velocipede'. That was 1861.

The machine moved forward by only the circumference of the front wheel at every turn of the pedals, so manufacturers made the front wheel larger and larger, producing the 'Penny-farthing'. With a very large front wheel, the rider perched on top and a small trailing wheel for balance, although these were easy to ride, tumbles were common. Yet a bicycling craze began; clubs were set up in the 1880s and one Thomas Stevens rode such a bicycle across the USA.

The propulsion breakthrough also came in the 1880s, with the first 'bicycle chain' with metal links to drive the back wheel from pedals. Different-sized gearwheels had the back wheel turning faster than the pedals; a 'freewheeling' mechanism let the rider stop peddling without stopping moving.

With the addition of inflatable rubber tyres (p. 105) in 1888, ball bearings to reduce friction in the works and many experiments with the form of the frame—ultimately made out of steel tubes—bicycles such as John Stanley's 'Rover' gained the form we see today, and unprecedented popularity. People now had the speed and freedom of movement they once enjoyed riding horses, and they (and the better roads bicycling demanded) were ready when bicycles were overtaken, both literally and figuratively, by the next big thing in personal transport, the automobile (p. 146).

The First Synthetics

Alexander Parkes, the Hyatt Brothers

The first synthetic materials were developed in the mid-nineteenth century, challenging some long-used 'natural' or 'organic' materials. Really they were only chemical manipulations of natural substances. For example, 'Vulcanite' was first produced around 1843 by heating natural rubber with sulfur (p. 89). It was hard and could be shaped but was always black, unless painted.

Prolific English chemist Alexander Parkes (80 patents plus 20 children by two wives) sought to improve Vulcanite by making it able to be coloured or transparent. By 1862 he had his wonder material, soon dubbed 'Parkesine'; it won a medal at the International Exhibition in London.

Parkes started with a known compound, cellulose nitrate, made by reacting cellulose, found in cotton, with nitric acid. Too much nitric acid produced explosive 'guncotton', a valuable gun-powder substitute (p. 127). When made with lesser degrees of 'nitration', cellulose nitrate could be dissolved in a solvent like ether to become 'colloidin'. Invented around 1848, this had a multitude of uses—in photographic plates, shatterproof glass and as a dressing for wounds.

When the solvent evaporated, a hard, waterproof and elastic residue remained.

Parkes built on that. Cellulose nitrate was dissolved in a minimum of solvent, heated and rolled, coloured and pressed into shape as needed. Parkesine was the first 'thermoplastic'; it softens when heated.

Parkes tried to market his invention but could not balance quality against cost. He claimed it was cheaper than Vulcanite, but that was possible only if the quality was poor, and it would not sell unless it was cheap. His company failed in 1868.

It fell to the American Hyatt brothers, John and Isaiah to succeed, a few years later. Their secret was to use camphor (from the laurel tree) as the 'plasticiser' in cellulose nitrate, something Parkes had tried. This produced a synthetic ivory, and 'celluloid', as the material was called, was first used in billiard balls. Billiards was very popular and thousands of elephants had been hunted to supply enough balls. Celluloid slowed the slaughter.

Celluloid was the first mass-produced plastic; it was used in toys, knife handles, collars and cuffs, photographic film and many other products. Around the middle of the twentieth century, due to its flammability and tendency to decompose spontaneously, it was replaced by safer, better materials like cellulose acetate and polythene. But the first synthetics had certainly shown what they could do.

Kicking a Ball Around

The Many Forms of Football

Games where a round ball (originally an inflated pig's bladder) is kicked about by two teams date back to ancient China and Rome at least. The rules have varied widely with time, place and circumstance. Modern codes arose within a few decades in the late nineteenth century from efforts to standardise rules for games in schools and elsewhere.

Rugby vs Soccer

According to the popular story, Rugby school student William Webb Ellis 'took the ball in his arms and ran with it' in 1823, defying the local rules. This probably apocryphal act was not reported for 50 years. In any case the rules were constantly changing, so Ellis may not have acted illegally, but it seems that old boys from Rugby popularised the 'carrying game', founding the first football club in the world at Guy's Hospital in London in 1843.

The key date in the evolution of football is 1863. The growth of football beyond schools led to the founding in that year of the Football Association (FA) at the Freemasons Tavern in London (the term 'association football' has now been shortened to 'soccer'). Much argument ensued over the rules, which still varied from club to club, particularly whether a player could catch the ball and run with it, and whether opposing players could try to stop him by holding, tripping or kicking him in the shins.

The FA decided to ban such practices and stick strictly to a kicking game. A number of clubs promptly withdrew, led by the Blackheath Club, whose representative complained that the 'pluck and courage' would be drained from the game. In 1871 those clubs met at the Pall Mall Restaurant to found the 'Rugby Union'. Soon after, England and Scotland met in the first rugby international, and the divorce between the two forms of the game was complete. Nowadays only the goalkeeper in a soccer game may catch the ball and there are strict rules on tackling.

In early rugby games, the objective remained to kick a goal during play but carrying the ball beyond the goal line entitled the attacking team to 'try' to kick a goal unimpeded, hence the nomenclature. Over time, 'tries' became more valuable than goals.

Technology made its impact. 'Pig skins' were replaced by bladders made from the new 'vulcanised' rubber (p. 89), enclosed in leather for durability. Balls remained round for the kicking game but were elongated for the carrying game, where the ball was passed hand-to-hand.

Union vs League

Rugby was sundered into two codes with different governing bodies and some distinctive rules in 1895 (and in Australia in 1908) driven by the growing practice of some clubs, especially in the north of England, to pay their players. For the next century, 'rugby union' remained an amateur (and therefore mostly middle-class) game while 'rugby league' became professional and popular among the working classes. Some of that atmosphere remains, though elite rugby union players are now remunerated.

Grid Iron

American football (often dubbed 'grid iron' from the complex array of lines marking the field) began as rugby, which was well established in the leading colleges (the 'Ivy League') of the north-eastern USA by the 1870s. Two differences emerged early—greater importance given to the try ('touchdown') in the scoring, and sides limited to 11 players (following the practice at Eton in England).

Innovations soon transformed the game. Rule changes in 1880 gave possession of the ball to one team, which was free to play it forward from the 'line of scrimmage'. To enliven the game, the team with the ball became required to carry the ball forward at least 5 yards (4.5 metres) in three attempts ('downs') or lose it; nowadays four downs are allowed to advance 10 yards (9 metres).

From 1906 players could pass the ball forward (strictly forbidden in the rugby codes) and in time the ball was reshaped to augment the necessary aerodynamic qualities. From early days, the ball carrier could be shielded from defenders by other players running in front; the notorious 'Flying Wedge' was invented at Harvard in 1892 and caused many injuries. Players not carrying the ball could be tackled (also not allowed in rugby). Solid body contact through blocking or tackling remains a feature; players are protected by helmets and padding. Over time, liberalised substitution rules encouraged extreme specialisation, and development of separate offensive and defensive teams.

Football Around the World

Soccer is now the 'world game', played in 200 countries, largely as a consequence of colonialism and the presence of British professionals in developing countries. The American game dominates in the USA; rugby codes are popular in a number of countries. Other variations of football include the Gaelic code in Ireland and the not-unrelated Australian Rules, devised during the late nineteenth century to keep cricketers fit in winter. Played on large grounds, these feature a combination of kicking, carrying and passing.

1863
Visions of the Future
Jules Verne, H.G. Wells and Others

Devising new technology is a creative endeavour, as much a work of the imagination as music, literature, painting or even science. Insight, as much as technical skill, is needed to develop a solution to a problem, or even to realise a problem exists. Many inventions have been conceived long before they could be realised, as the visions of da Vinci (p. 12) and Francis Bacon (p. 25) attest. While mostly concerned to tell a good story or introduce some interesting characters, works of fiction often contain descriptions of inventions not yet made or even attempted, or farsighted elaborations of technologies already available.

The nineteenth century saw outstanding examples of such imagination, in books we would now call 'science-fiction' ('scientific romances' was a more contemporary term). In the vanguard was French writer Jules Verne, some of whose novels were simply adventures, others show insight into the future of technology. *Five Weeks in a Balloon*, his first such book published in 1863, and *Around the World in 80 Days* pushed the transportation technology of the day towards its limits. Captain Nemo's submarine in *20,000 Leagues Beneath the Sea* was based on the existing concepts of Robert Fulton and gave the vessel the same name, though with greatly enhanced capabilities.

In *From the Earth to the Moon*, Verne prophesied space travel even if he was not the first to do so and his launch technology simply a scaled up gun, driven by very familiar gunpowder. Other works gave foretastes of technologies we have since found familiar; helicopters, artificial satellites, rocket propelled missiles and even nuclear-power.

A generation later British writer H.G. Wells, who had training in science and much experience as a science journalist, pushed the envelope. We are yet to see anything comparable to his *Time Machine* (1894), the hybrid creatures created on *The Island of Dr Moreau*, the antigravity 'cavorite' that made possible *The First Men in the Moon* or the technology to create *The Invisible Man*. Yet the clouds of poison gas used by the invading Martians in *The War of the Worlds* had their counterparts in the coming Great War, their heat rays find parallel in today's laser beams and we are no longer amazed at the scale and power of their walking machines.

Like Verne, Wells foresaw space flight, though not by rocket power. *When the Sleeper Wakes* incorporates powered flight, television and something akin to radar years or decades before they were even embryonically real. In his 1914 *The World Set Free*, a chemist liberates nuclear energy a quarter-century before nuclear fission was known. These are part of broader theme. Both the *Time Machine* and *The Sleeper Wakes* are predicated on the future being very different to the present or the past, at least in terms of technology; *The War of the Worlds* allows civilisations existing at the same time to have very different technological capacities.

Later, in Wells' increasingly pessimistic mind, such technological advances became vehicles of degradation and destruction, enslavement rather than liberation. His early writings hint of this. *The War in the Air* describes the potential of the newly invented aeroplane to inflict destruction in wartime, a prophecy made only five years after the triumph at Kitty Hawk and fulfilled when the Great War erupted half a decade later. Well's deeper message was that evermore powerful military technology had made the existing international order based on nation states unviable and some supra-national government must take its place to ensure human survival.

Sombre messages continued to multiply in the twentieth century, paralleling a loss of confidence in science. George Orwell's chilling masterpiece *1984* may have been mostly an indictment of the totalitarian government of his day, but the exploitation of technology adds bite.

Television, already in use in 1948 when he was writing, becomes a mechanism for state intimidation and mind control; Winston Smith's 'speak-write' dictating machine a simple means to continually recreate the past.

Aldous Huxley's apparently carefree *Brave New World* is totally shaped by technology, from the vast human reproduction programs and sleep conditioning used to generate a stable stratified society, to the universally available 'Soma' mind-soothing drug and games like 'elevator squash' and 'electromagnetic golf' which epitomise conspicuous consumption. As the conversation between the Savage and the Controller shows, the quest for security through technology can mean sacrificing much of what it means to be human.

Closer to our present time, hundreds of 'science fiction' writers are at work, some much more insightful than others (as in any genre), and many alternative visions of tomorrow are available. Some ignore the Laws of Nature as they are currently understood by science, and are unlikely to be fulfilled. But science fiction in one decade can come close to reality in another. Devotees of the classic TV series 'Star Trek' like to point our how close the 'communicators' seen in early series resemble today's mobile phones.

Get Your Skates On

James Plimpton, the Olsen Brothers

Skating on ice has a long history in cold climates where lakes and rivers freeze over. Its origins are lost in time. More recently, adventurous people tried replacing the blades with rows of wheels so you could skate on a smooth, non-icy surface. One of these, according to a possibly apocryphal story, was London-based inventor Joseph Merlin. In 1760 he wanted to make a grand entrance at a party by smoothly gliding in playing his violin. Sadly his 'in-line' skates were very hard to steer and he came to grief before the packed ballroom.

What we call the roller-skate today was invented by American James Plimpton in 1863. Seeking both stability and manoeuvrability, he used two pairs of wheels, one under the heels and another under the instep (the precise name is 'quad skates'). For the first time, skaters on wheels could move in a smooth curve and even skate backwards. The popularity of roller-skating began to boom. The use in the 1880s of ball bearings, then being incorporated into bicycles (p. 117), made skating still easier and skates lighter. In the 1960s tough plastic wheels (such as polyurethane) replaced wood and rubber. Over the century the popularity of roller-skating waxed and waned, from frenzied to all-but-forgotten.

The venerable in-line skate, as used by Merlin, was revived by the ice hockey-playing Olsen brothers of Minnesota. They came across an antique pair of pre-Plimpton skates. Thinking initially of providing off-ice training for their team, they incorporated modern materials and ideas from quad skates, and 'roller-blading' was born. Their first model was launched in 1983, but it had a number of technical problems and the Olsens soon sold Rollerblade Inc to someone better equipped to make a go of the innovation. New materials and better design reduced the weight and increased safety. Within a few years, sales were enormous.

Meanwhile the quad skate had morphed into something new, the skateboard. Pairs of roller-skate wheels were attached under the ends of a short plank. It appears this originated in the 1950s among Californian surfers who needed something to do when the waves and weather would not cooperate. Early models were hand built. Devotees soon found that the skateboard needed the same 'oneness' with the board as surfing did. New materials and better technology brought the capacity for trick riding and cemented the skateboard as a key component of youth culture.

Pasteurisation Prolongs Shelf Life

Louis Pasteur

Louis Pasteur carries one of the greatest names and reputations in the history of science. Through the late nineteenth century, this French biologist won the battle on behalf of his 'germ theory' against its rivals. He maintained, and then conclusively proved, that plant and animal diseases are caused by microscopic living things, 'microorganisms' or 'bacteria'. The same is true of the fermentation of wine and beer, the souring of milk, the decay and corruption of all kinds of once-living materials and even the digestion of foods in the intestines of animals. All these are the consequences of chemical changes induced by the microorganisms as they release the energy they need to live.

This new understanding, which included the revelation that many microorganisms do not need oxygen to live, had profound impacts on technology. It transformed medicine by generating the new 'antiseptic' or 'aseptic' surgery of Joseph Lister (p. 126) that saved many lives by reducing deadly infection in surgical wounds. It led Pasteur himself to develop 'vaccination' against many diseases (p. 143). In industry, it enabled the makers of products based on fermentation, such as alcohol, to enhance quality by ensuring only desirable microbes were present, since each microorganism catalyses only one particular reaction.

Pasteur's name is also recalled when we refer to pasteurisation, widely used for the last 150 years to make foods last longer and to stop fruit juices and dairy products going sour. Its origins lay in Pasteur's discovery that the responsible bacteria are sensitive to heat and certain chemicals, and so can be killed or incapacitated. In 1864 Pasteur was asked by Napoleon III to investigate diseases threatening to ruin the wine industry. He found that heating the wine to 55°C for a few minutes disabled the bacteria and prevented the disease. Similar treatments to beer and milk followed soon after and 'pasteurisation' entered the language.

Generally, the higher temperature used, the less heating time is needed. Pasteurisation of milk now takes place continuously rather than in batches. Milk is judged pasteurised after being heated to 62°C for 20 seconds, but it is not completely sterile and still requires refrigeration. Ultrahigh-temperature (UHT) pasteurisation, used on fruit juices and 'long-life' milk, which then require no refrigeration or preservatives, involves heating to over the boiling point, but under pressure so no boiling actually occurs. This can affect flavour, and all pasteurisation reduces the nutritional quality of foods.

Cement
United with Iron

Joseph Monier, Francois Hennibique

By the mid-nineteenth century cement (p. 49) was a popular building material, increasingly replacing wood and stone. But it had problems. Although cement has great compressive strength—it bears loads well and is hard to crush—it is weak under tension. Try to stretch it, and it easily cracks and crumbles.

This challenge was addressed by many, most notably by French commercial gardener Joseph Monier. He pioneered the use of steel mesh embedded in concrete tubs and basins, the kind used in horticulture, patenting the idea in 1867 and displaying it in the Great Paris Exposition in that year. Monier had no technical training in these matters, but seemed instinctively to know what he was doing.

Over the next decade, he extended his concept of 'reinforced cement' (and the number of his patents) to cement pipes, water tanks, reservoirs, panels for building facades, foot and vehicle bridges, the sleepers or ties supporting railway tracks and concrete beams for structural strength within buildings. The embedded wire took the tensile stresses, the cement handled the competitive forces, the two components functioning as a unit. The range of uses to which cement and concrete could be put was vastly expanded. A bridge built with reinforced concrete opened in San Francisco's Golden Gate Park in 1889; it is still in use. Within a few more decades, reinforced concrete would be forming the huge locks in the Panama Canal.

Among those who admired Monier's innovation on display at the Paris Exposition was engineer François Hennebique. Quick to see the possibilities, he set to work to devise ways to apply the technology to buildings, using steel rods for reinforcement. By 1879 he was making whole floor slabs from reinforced concrete, and by 1892 had devised and patented a complete building system, using reinforced concrete floors, beams and columns, connected by metal stirrups. He used this to construct an apartment building in Paris, setting a pattern for building construction that continues largely to the present day.

Buildings were now on the way up, supported by reinforced concrete and ever-improving steel in its core and frame, rather than by massive outer walls. Soon would come buildings so high they would be termed 'skyscrapers', be they only 20 stories. And something else liberated buildings to reach for the sky, the new 'elevator' invented by Elisha Otis (p. 111).

Surgery without Infection

Joseph Lister

Two hundred years ago, surgery was agony for those under the knife, and often ended in death as surgical wounds became infected. Two developments changed all that: the use of anaesthetics to control pain (p. 107) and the introduction of antiseptic surgery. The great pioneer in the second revolution was English doctor Joseph Lister.

Lister's family were prosperous Quakers; he became professor of surgery at the University of Glasgow. The all-too frequent and often deadly infection of surgical wounds was a challenge to all surgeons. Patients often died after otherwise successful surgery. At that time, septic wounds were blamed on 'miasmas' of bad air. Surgeons saw no reason to wash their hands before an operation; their bloodstained operating coats were never cleaned. Bed linen was rarely changed and usually dirty. The appalling stench of rotting wounds filled the wards.

Lister developed an alternative explanation for wound infection, based on the 'germ theory' pioneered by Louis Pasteur (p. 124). The Frenchman had shown that fermentation and putrefaction was caused by microorganisms, active even if deprived of oxygen. Lister confirmed this with the microbes he found responsible for the often-deadly infection gangrene.

To kill microbes, Pasteur had suggested heating, boiling or exposure to chemicals. Lister began to experiment with carbolic acid (phenol), distilled from coal tar and known to remove the smell from rotting sewage, as an 'antiseptic'. His methods were comprehensive. Swabs for cleaning wounds and bandages for covering them later were soaked in a solution of carbolic acid, the surgical instruments disinfected and the surgeons' hands similarly washed. The air around the operating table was sprayed with carbolic vapour.

Though the use of the irritating spray was later shown to be unnecessary, Lister obtained impressive results. Where carbolic was used, gangrene rates dropped markedly; many more patients lived.

In 1867 Lister was able to announce 'The Principle of the Antiseptic Practice of Surgery' in the leading medical journal *The Lancet*.

Despite much scepticism, indeed scorn and ridicule, among his colleagues, Lister had been on the right track, and antiseptic or 'aseptic' surgery is now universal, though with more effective and convenient agents than Lister's carbolic.

The shy, deeply religious Lister became a Baron and one of the first members of the Order of Merit. He is also remembered through the antiseptic mouthwash Listerine, and the microorganism *Listeria*, a common cause of food poisoning.

Bigger, Safer Explosions

Christian Schönbein, Alfred Nobel

Nineteenth-century miners hunting for coal and metal ores, and engineers building canals, bridges and railroads often needed to blast to break up rock. However, they had only one explosive to work with, the venerable gunpowder—a mixture of sulfur, carbon (charcoal) and nitre (salpetre). The last of these supplied the oxygen in which the others burnt rapidly, liberating the very hot, high-pressure gases that generated the explosive force. As its name implies, the first (and still main) use of the explosive was in guns, both cannons and small arms.

The first alternative explosive was invented, apparently by accident, by the Swiss chemist Christian Schönbein (who also discovered ozone). In 1845, the story goes, he was experimenting at home and spilt a mixture of nitric and sulfuric acids. He mopped the mess up with his wife's cotton apron and hung it up again. When this was dry, it suddenly vanished in a puff of smoke, completely consumed.

The mixture of acids had added small clusters of atoms called nitrate groups to the cotton, making 'nitrocellulose'. These groups supplied oxygen for rapid burning just like the salpetre (potassium nitrate) in the gunpowder. The new explosive was first used in warfare, hence its name 'gun-cotton'. Its white smoke did not obscure the battlefield, unlike the black smoke of gunpowder.

Chemists were soon adding nitrate groups to other flammable substances, seeking new explosives. The most powerful and hazardous was nitroglycerin, discovered in 1847 by the Italian chemist Ascanio Sobrero. This was too unstable for use in war; even peacetime use in civil engineering required very careful handling. The slightest shock could set it off; carelessness cost many lives.

The Nobel family from Sweden was a major manufacturer of the explosive (they later employed Sobrero). When his brother was killed in a factory explosion, Alfred Nobel worked to make nitroglycerin safer. In 1867 he soaked an absorbent mineral in liquid nitroglycerin and produced a stable material, able to be moulded into sticks for easy handling: dynamite. Combining high-explosive force with acceptable safety, it became tremendously popular in both peace and war.

Together with gelignite, a mixture of nitroglycerin and guncotton, the explosive funded the Nobel Prizes, first awarded in 1901.

The hunt for better explosives went on. Early in the twentieth century, three nitrate groups were added to molecules of the hydrocarbon toluene, so making the widely used trinitrotoluene or TNT.

Machines that Write

Christopher Scholes

The first reference to an 'artificial machine for the impressing of letters', today called a typewriter, is nearly 300 years old. England's Queen Anne granted a patent to one Henry Mill in 1714, the patent document claiming the results 'could not be distinguished from print'. Nor could letters 'be erased or counterfeited without manifest discovery'. Sadly we do not know how his machine worked, or if it worked at all.

A hundred years later, the US government granted a patent for a 'typographer' to William Burt from Detroit. He did not get very far with it, especially once the only model was lost when the Patent Office burnt down in 1838. By all reports, we did not lose much either; the typographer was so clunky anyone could write faster than they could type with it.

Another quarter century passed before a practical typewriter emerged, this time from the American pair Christopher Scholes and Carlos Glidden. These two tinkerers frequented Kleinstueber's Machine Shop in Milwaukee, Wisconsin, hoping to come up with inventions to help mankind and perhaps enrich themselves.

In the first model, around 1868, pressing on old telegraph keys banged a piece of type through carbon paper to mark a page. There was no way to advance the line or provide a space, but it was a start.

Five years of work led to two patents and a machine good enough to sell. Scholes was no businessman. He sold his rights to his financial backer James Densmore, who then induced the rifle manufacturer Philo Remington to build and market the machines. Hence the Remington name is more famous in typewriter history than Scholes or Glidden.

The first model sold less than 5000 machines, but within a few years improvements by the Remington engineers, experienced in mass-producing machinery, made the typewriter much more useful. These included the 'shift key' to enable type in both upper and lower case. Sales soared. The lives of office workers were changed forever: new careers opened up for women, who would make the office typewriter their own for the next century.

Over that time typewriters continued to advance, especially once they became electrified. In recent years the typewriter has become the word processor and then the 'keyboard', the universal interface with computers and the Internet.

The first author to submit a novel in typescript was Mark Twain; the book was *Life on the Mississippi.*

Air Brakes and Electricity

George Westinghouse

In 1866 the train carrying 20-year-old New Yorker George Westinghouse, recently returned from the Civil War, narrowly avoided running into a train wreck. With the brakes applied by hand, stopping had taken hundreds of metres.

The inventive Westinghouse already held a patent for a rotary steam engine. Wanting to stop trains more quickly and safely, he experimented with steam power, and then with compressed air, connecting carriages with a flexible pipe so all brakes could be applied from one location. Patented in 1869, the Westinghouse air brake was 'failsafe' once fully developed. Air pressure was needed to release the brakes, not to apply them. Loss of pressure (for example, by a carriage uncoupling from the rest and breaking the line) automatically applied the brakes.

In 1893 automatic braking systems in trains were mandated by law. By 1905, 90 000 locomotives and two million cars and carriages in the USA carried the Westinghouse system. Trains could now travel safely at higher speeds, reducing travel times, increasing profitability and transforming the industry. Safety was further enhanced by his 1880s invention of automatic train signals that warned of blockages ahead.

Westinghouse now had the resources to become active across a wide range of manufacturing industries. Moving into electricity generation and supply brought rivalry with Thomas Edison (p. 151), but his advocacy of alternating current (AC) over the direct current (DC) Edison favoured was to win out in the end, especially through his collaboration with the highly creative Nikola Tesla (p. 145), master of the technology of AC generators and motors. He also boosted the feasibility of electric trains by powering them with a new AC motor, first demonstrated in Pittsburgh in 1905.

When he died in 1914, Westinghouse held 390 patents, including a method for the safe transmission of natural gas into homes and factories, and an early form of telephone exchange. His last patent described a compressed air spring to smooth the ride in the new automobiles. He also held exclusive rights to manufacture the steam turbine invented by Charles Parsons (p. 142).

Westinghouse innovated in business too, being among the first to give employees a half day off on Saturdays, paid vacations and access to a pension fund. The company bearing his name (which passed from his control in the financial panic of 1907) is a global player today. It owned KDKA, the world's first radio station, which went on air in Pittsburg in 1920 (p. 189).

The Staple, the Paperclip and the Safety Pin

George McGill, Johan Vaaler, William Hunt

How you hold two pieces of paper together so they won't get separated (unless you want to separate them)? This simple question, a vital one in our paper-driven age, has produced many answers: passing string or ribbon through holes or slots cut in the pages (time-consuming), or pushing a straight pin back and forth through the leaves (not very secure). Better solutions included brass fasteners with two legs passed through a hole and spread out on the other side. Two more recent responses are the staple and the paperclip—simple, even obvious, but needing to be invented.

American inventor George McGill (already the inventor of the brass fastener and its associated hole punch) devised the staple around 1870, though others probably had a similar idea. A U-shaped piece of fine wire was pushed through the pages and the free ends flattened back to hold them together. Now very familiar, then very new.

Early staples were fed into the machine one by one or formed from a long piece of wire. Only around 1895 did we get the now familiar strip of staples connected together, struck off individually and punched in as needed. 'Stapling' (the term was not used till around 1900) became popular, particularly as a way of holding the pages of a magazine together along the spine, and demolished most of its competitors.

Coming close behind was the paperclip, which did not make holes in the paper and was very easy to remove. Norwegian inventor Johan Vaaler usually gets the credit; certainly he secured the first patent. It is not Vaaler's 1899 design we recognise now, rather the double oval 'Gem' model, devised but never patented by an English firm, also in 1899. The extra loop prevents the steel point scraping across the page.

We can update the story of the straight pin. This 4000-year-old invention has well-known problems; it slips out and can prick your fingers (or the baby). The solution probably came in stages. Someone bent a pin double so that its head and point could be latched together. Then in 1849 prolific American inventor William Hunt wound the pin into a spiral spring, exerting pressure on the latch to hold it closed. Hunt reportedly sold his newly patented idea, perfected in three hours of wire bending, to a friend for $400 to settle a small debt. Nowadays the safety pin usually holds other things together, rather than sheets of paper.

The 'Devil's Rope': Barbed Wire

Joseph Glidden

Anyone could have invented barbed wire. Putting points onto smooth fencing wire was an obvious thing to do, and many designs were tried during the late nineteenth century. Demand was growing, especially out on the vast American prairies then being settled, and where, according to some accounts, fencing cost as much as livestock.

Fame and fortune therefore went to the man who made the idea work, devising machinery to make the stuff by the kilometre. At an a Illinois country fair in 1873, 60-year-old rancher Joseph Glidden saw a new form of fencing, merely a wooden rail with protruding metal spikes hanging behind a smooth wire fence. Glidden thought he could do better by fixing the spikes to the wire. His first lengths of hand-made barbed wire surrounded his wife's vegetable patch, and he had thoughts on mechanised manufacture. His friends Jacob Haish and Isaac Elswood, having seen the same demonstration, had similar plans. Glidden secured a patent in 1874, and then spent several years in court battling Haish, Elswood and others over ownership of the intellectual property. He emerged victorious as the 'father' of barbed wire.

Glidden, a teacher from New Hampshire before turning to ranching, made a fortune from royalties on his invention, dying in 1906 aged 93 as one of the richest men in America. Sales boomed, as barbed wire remade the West. Elswood was making his own pile from manufacturing the wire; output from his factories went from 1000 tonnes in 1876 to 25 000 tonnes in 1879.

Success was not inevitable. 'Free-rangers' were against any sort of land enclosure, and religious groups called barbed wire the 'Devil's Rope' because it could injure animals encountering it for the first time. Texas ranchers, a large potential market, doubted the flimsy-looking fence would contain their powerful longhorn cattle. A public demonstration in 1881 changed many minds. Wild cattle were driven into a barbed wire enclosure. Agitated by the onlookers, the cattle repeatedly charged the fences, but they held and the animals suffered little harm. Soon barbed wire fences were criss-crossing the West, including along the rapidly spreading railroad tracks (p. 92) to stop farm animals from straying onto the line.

Barbed wire had a more macabre application a few decades later, erected along hundreds of kilometres of trench lines in World War I. Many soldiers on both sides met their deaths tangled in the 'Devil's Rope'.

Cleaner Carpets, Cleaner Floors

Melville Bissell, James Spangler, William Hoover, Herbert Booth, James Dyson

Melville Bissell ran a china shop with his wife Anna in Grand Rapids, Michigan. Allergic to the dusty straw the china came packed in, he was driven to invent a 'carpet sweeper' in 1876. Rotating brushes gathered dust and other rubbish into a metal box, pushed by a long handle. It sounds very familiar. The components were initially made by Bissell's friends and neighbours in their homes. Much easier than beating the carpet out-of-doors, Bissell's sweeper immediately became widely popular. Housewives and housemaids spoke of 'bisselling the carpet'; the next-generation would talk of 'hoovering'.

Ohio businessman William Hoover bought the patent for the upright 'vacuum cleaner' from his wife's cousin James Spangler, a department store janitor. Spangler's own needs, like Bissell's, drove his inventiveness. Blaming dust for his chronic cough, he cobbled together the electric motor from a fan, a soapbox, a pillowcase to catch the sucked-in dust and the obligatory broom handle. Hoover's first 'bag on a stick' machines, marketed in 1907, were little better than Spangler's prototype, but they sold briskly, with Hoover offering a 10 day in-house trial. Sales soon went worldwide.

Across the Atlantic in Britain, Hoover's rival was Herbert Booth (or Bothe), who marketed cleaners under the Goblin brand. Booth may have been first to think of sucking up the dust; previous air-powered machines such as the 1869 Whirlwind blew it away. His first machines were huge, carried around the streets in carts during 'spring-cleaning' time. Long flexible pipes connected to cleaning heads were passed through windows to 'vacuum' the carpets and floors. Booth's contraption cleaned the Westminster Abbey carpet for the coronation of Edward VII. The King was impressed and bought two, but most people preferred to own smaller machines. Booth obliged.

Beyond the disposable paper dust-bag, invented in 1920, vacuum cleaners changed little until the 1980s. Then James Dyson conquered the big defect of traditional cleaners; air is drawn through the dust bag and so 'suction' weakens as the bag fills. Dyson's 'cyclone' technology dispensed with the bag, using a centrifugal device to separate air from dust, collecting the latter in a cylinder. Initially spurned by manufacturers, Dyson's machines claimed 30 per cent of the market by volume by 2001 and 50 per cent by value.

The next fashion may well be robot cleaners able to vacuum unattended; they are already on the market.

The Birth of the Telephone

1876

Alexander Graham Bell, Elisha Gray

On 14 February 1876, lawyers for both Alexander Graham Bell and Elisha Gray contacted the US Patents Office within a few hours of each other. Bell's lawyer lodged a patent; Gray's gave notice that one would be lodged within three months. Their clients had invented, simultaneously and independently, devices to send sounds, including speech, as an electric current on a wire. For reasons of priority, Bell got the patent; Gray did not, despite the backing of the telegraph company Western Union. So Bell is the name we remember.

The two men had very different backgrounds. Bell was the Scottish-born son of a university professor, trained in medicine and with expertise working with deaf children. Gray was a Quaker from rural Ohio, a decade older than Bell, and had worked as a blacksmith, boat builder and carpenter, later paying his way through college to win a two-year science degree.

Both had worked to improve the operation of the 40-year-old electric telegraph, such as developing ways to send more than one message at a time. Both had devised systems that made an electric current vary in synchrony with the minute changes in air pressure that constitutes sound. In B ell's design, a metal diaphragm responded to the pressure changes and moved a coil of wire near a magnet. The movements 'induced' a varying electric current (p. 95). At the receiver, those changing currents powered an electromagnet (p. 88), which pulled a second diaphragm to and fro, simulating the original sound. That is how Bell's assistant, the electrician Thomas Watson, heard the boss calling from several rooms away, 'Mr Watson, come here, I need to see you', not through the air but over the wire.

Bell and Gray were not alone in this field. The German Johann Philipp Reis, among others, had already built primitive devices; still others would make improvements on Bell's design. In 1877 Thomas Edison replaced the coil-and-magnet receiver with one containing a button of carbon, improving fidelity. Even so, the telephone was not quite ready for mass use. Over the next few years, Bell, Watson, and others added a hand crank to signal down the line that a call was being made (by ringing a bell), and put the receiver and transmitter on a common handle. This hand-piece hung on a hook; lifting it opened the line to the 'exchange'. There, 'operators', later mostly women, put plugs into sockets to connect callers to their desired destinations.

The first exchange began operating in 1877. 'Dial phones' instructing automatic exchanges were still 40 years away.

The appeal of the telephone over the telegraph was its immediacy, convenience and privacy. You heard the voice of the person you are talking to, rather than just reading their words. Messages could be sent from your own home, rather than needing a trip to the telegraph office. Once the call was established between the two parties, no one else need interfere.

Bell's telephone caused a sensation at the 1876 Centennial Exposition in Philadelphia; the Emperor of Brazil reputedly put a receiver to his ear and dropped it with a start, crying 'It talks!' The phone was not, by today's standards, an overnight commercial success. In 1895, 20 years after its invention, less than 300 000 were in use in the USA, about one in 50 of the population. Long-distance calls, possible though 'electronics' (p. 176), made the telephone more popular; numbers trebled in the next 20 years. By the 1950s, just about every home in an industrialised nation had one, though a threat to the dominance of the 'landline' would emerge a few decades later from the 'cell-phone' (p. 275).

Despite losing the patent battle over the telephone, Elisha Gray did not fade into obscurity. He became a well-regarded university professor, and held 70 patents.

His company, Western Electric Manufacturing, became today's Lucent Technologies. A deeply religious man, he wrote extensively on the nexus between science and faith, and advocated the concept of 'intelligent design', much discussed today. He died in 1901.

Aged only 30, Bell was rich and famous, married to the daughter of his financial backer, a Boston lawyer keen to break the Western Union monopoly. With his Bell Telephone Companies rolling out wires on poles throughout cities and towns to link his telephones, Bell continued to innovate. His mind remained active until his death at 75. Among his many creations, he thought most highly of his 'photophone', which sent sounds on a modulated beam of light, anticipating today's 'optic fibres' (p. 262). The death of his newborn son from respiratory problems led him to construct a metal jacket to aid breathing. This presaged the 'iron lung' used by polio victims many decades later. Water desalination, hydrofoils, sheep breeding to increase multiple births—all these caught his attention; he was also an early and passionate pioneer of powered flight (p. 172).

Recording Sound

Thomas Edison, Emile Berliner

In 1877 Thomas Edison, trying to improve telegraph operations, built a machine that marked the dots and dashes of morse code as indentations on paper tape, ready to be sent through a reading device. As the paper passed through the machine at high speed, the spring on the stylus reading the impressions made a musical sound. Edison grasped that such indentations could record and playback sounds by connecting the stylus to a diaphragm, as in the newly invented telephone (p. 133).

The prototype 'phonograph' ('writing in sound', analogous to 'photograph') used metal foil around a cylinder to carry the indentations. The first words ever recorded were of 'Mary had a little lamb'. Soon after, Edison demonstrated his new machine to the staff of *Scientific American,* having it wish them 'goodnight'.

He started selling the machines the next year, but foil cylinders were awkward and fragile and interest in the novelty soon waned. Before moving to other pressing matters such as the light bulb (p. 137), the visionary Edison listed potential uses for the phonograph: 'talking books' for the blind, clocks that told the time (over the telephone), reproducing music, recording telephone messages, taking dictation and preserving the last words of the dying.

Coming back to the phonograph 10 years later, he found that others, like telephone pioneer Alexander Bell, had made improvements, using wax instead of foil for the recording. Machines became 'spring powered', rather than turned by hand. Edison soon had a range of pre-recorded cylinders for sale: sentimental ballads, marches, comic monologues and opera excerpts. But recordings were limited to two minutes in most cases, and there was difficulty in reproducing cylinders in large numbers.

By the time of World War I, the Edison cylinder had given way to disc recordings. German-born Emile Berliner, who had emigrated to Washington DC aged 19, had devised a technique in 1887 to record sound in a spiral groove on a flat disc. The impressions coding the sound went side to side rather than up and down as in the Edison machines, making for better reproduction. The recordings could run four minutes, and most importantly, could be mass-produced by stamping an etched metal 'master disc' onto blank discs of shellac softened by heating. His name 'gramophone' reversed Edison's title. His system, with modifications, underlay all recorded sound for more than half a century, until challenged by sound on film (p. 194), tape recordings (p. 162) and now compact discs (p. 287).

Separating the Cream from the Milk

Gustaf de Laval

Oil and water don't mix. Farmers used that fact for centuries to separate oily cream from watery milk. The cream is also 'lighter' (less dense) than the milk, floating to the top if given enough time, skimmed off the milk with a spoon or a special shallow dish (hence the term 'skim milk'). It was slow and not very efficient, a lot of cream remaining mixed with the milk.

Swedish inventor Gustaf de Laval heard that a German brewer had spun a barrel of milk to make the cream separate faster. When the German showed no interest in collaboration, Laval experimented alone. In his first machine, a number of buckets were chained to a vertical shaft. Spinning the shaft made the buckets swing out horizontally. The heavier milk particles were thrown outwards and ended up at the bottom of the bucket, leaving the cream behind, just as happened under gravity, but much faster. The 'centrifuge' (to use a later term) produced a form of 'artificial gravity'.

By 1878 Laval had a machine that separated continuously. Whole milk went in at 150 litres an hour; cream ('butter fat') and skim milk emerged from separate spouts. Dairy industry productivity rose, costs fell and the machines became very popular. Modern cream separators use the same principles but with greater efficiency. The separation of the high-value buttermilk from the less valuable skim milk is all but total.

Today centrifuges (the name means 'flying from the centre') are everywhere. Your spin-dryer (p. 109) is one, producing about 2000 times the force of gravity. Industry uses similar machines to separate solids from liquids, such as impurities from oils. High-speed centrifuges can separate the isotopes of uranium (p. 216) for power generation and weapons manufacture, purely on the basis of their slight mass difference. In medicine, centrifuges separate blood cells from the liquid (plasma) in which they are carried. 'Ultracentrifuges', which spin glass containers thousands of times a second with jets of air to produce a million times normal gravity, separate tiny particles of different proteins on the basis of their mass, a crucial capability in modern biotechnology (p. 273).

Though never rich, Laval was a prolific and much acknowledged inventor, making vital improvements to the steam turbine (p. 142), building a factory to make the new incandescent light globes (p. 137), experimenting with milking machines and showing great interest in early aviation.

Who Really Invented the Light Bulb?

Thomas Edison, Joseph Swan

Whatever the books say, Thomas Edison did not invent the electric light bulb. He certainly pioneered the mass distribution of electricity to light those bulbs. That undertaking needed much innovation; a network of buried copper wires, parallel circuits so that each bulb enjoyed the same voltage, sockets with on/off switches, safety fuses, more efficient dynamos (p. 95) to make the electricity and equipment to hold the voltage steady. Meters to measure how much power each customer bought came later.

Edison opened the first American generating plant in Pearl Street, Manhattan, in September 1882, supplying 50 customers and 1300 lamps. This was not the world's first power station; Edison had set one up in London nine months earlier. By 1890 many cities would be enjoying electricity from their own plants, with the Edison power companies transmuting into General Electric soon after. Edison could claim to have founded the electricity industry, though he shared that honour with George Westinghouse (p. 151), among others.

Credit for the light bulb as a concept belongs elsewhere, such as with Warren de la Rue in 1820. He made a platinum coil inside an evacuated glass bulb glow by carrying electricity from a battery, but the high cost of platinum precluded mass use. From the 1850s inventors in England and Germany experimented with glowing 'filaments' of carbon, made by charring bamboo or paper impregnated with tar. In 1878 English physicist Joseph Swan patented a light bulb that used a filament of blackened cotton. This stayed alight for 14 hours before burning out, but mass production of his lamps did not begin till 1881.

Edison was now busy too, exploiting a patent bought from Canadian inventors James Woodward and Matthew Evans. Their lamp had a shaped carbon rod held between metal struts in a hand-blown glass bulb filled with nitrogen to stop the filament burning away. Edison used his resources and ingenuity to the hunt for a cheap but durable filament. Carbonised cotton thread proved best. By 1879 his bulbs would burn for 40 hours, and by 1880, for 1200 hours, long enough to be commercial.

Modern light bulbs operate much as Edison's and Swan's did, though filaments are now made of tungsten, and the machine-blown bulbs are argon-filled. They remain inefficient; only 5 per cent of the electricity consumed by an electric lamp emerges as light, the rest being wasted as heat. Hence the importance of fluorescent lamps (p. 170) and LEDs (p. 256).

Measuring Earthquakes

John Milne, Charles Richter

Japan being one of the most earthquake-prone regions on Earth, it is not surprising that modern earthquake-recording machines or 'seismographs' were invented there. The major work was done, however, by three visiting professors from Britain: John Milne, James Ewing and Thomas Gray. Milne in particular urged the establishment of the Seismological Society of Japan in 1880, following a major earthquake in Yokohama. The society was the first ever devoted to the study of earthquakes.

Eighteen hundred years before, Chinese scientist Chang Heng had invented the first instrument to detect an earthquake. Many others followed, but these did not keep a continuous record of the shaking of the ground. In the Milne seismograph of 1880, the frame of the instrument carrying the recording drum was firmly fixed to the ground, shaking when it shook, underneath a heavy pendulum connected to a marking pen. When an earthquake occurred, the pendulum remained mostly stationary and the resulting trace registered how fast and how far the instrument (and therefore the ground) had moved, and the relevant time. With improvements in technology, this remains the basis of most earthquake recording today.

Milne was also the first to promote the establishment of networks of earthquake-recording stations. Comparing the various records, seismologists could determine the location and magnitude of the quake. Magnitude reports how much energy was released at the point of origin of the quake, how much the rocks moved there. American geologist Charles Richter invented a scale for this in 1935 (though he was always keen to credit the contribution of his colleague Beno Gutenberg). Initially it applied only to earthquakes near Los Angeles in southern California (where they are plentiful); Richter worked there. Now it is used around the world, and Richter's name is in our everyday language.

Earthquakes vary enormously in magnitude: destructive ones release millions of times more energy than small ones just large enough to be measured. So the Richter Scale is 'logarithmic'; each step represents a tenfold increase in energy released. A magnitude seven earthquake has ten times more destructive potential than a magnitude six, a hundred times more than a five, a thousand times more than a four.

The scale also reflects the likely occurrence of quakes of different magnitudes. On average, magnitude six quakes happen ten times more often than magnitude seven ones, 100 times more often than magnitude eight.

The Retailers Innovate

James Ritty, Montgomery Ward

Nineteenth-century mass production vastly increased the volume and range of manufactured goods for sale. Now innovation at the retail end of the chain could ensure those goods were efficiently exchanged for customers' cash.

Cash Registers

Dayton, Ohio, was the birthplace of one such innovation, the cash register. James Ritty was the youngest of five brothers, several of whom were inventive and held patents. He ran a saloon that seemed successful but constantly lost money. He was never sure how much, because no quick and easy record could be kept of the sales; money simply went in and out of the cash drawer behind the bar. He wondered if his barstaff, who turned over regularly, were helping themselves.

Broken by worries over his business, Ritty took a cruise to recover his health, during which he developed an interest in the ship's machinery. He soon noted a device counting the revolutions of the ship's propeller. Surely a similar account could be kept of the takings of a store. With his brother John, he built a succession of machines of increasing sophistication, which ultimately could show both clerk and customer what the goods had cost and print a record of the transaction on paper tape. Totalling the tape gave the day's takings.

'Ritty's Incorruptible Cashier' reached the market in 1881, but the brothers did not make a success of the business, selling out for $1000 after a few years. Others made the business work, and cash registers were soon ringing across America and elsewhere. Retail outlets became more efficient and profitable, owners and customers both benefited.

Mail-order Catalogues

In nineteenth-century America, as in all sparsely settled countries, not everyone had a convenient store nearby, not one selling goods like furniture or clothing or farm equipment. Many still lived on the land or in small rural hamlets. They came to rely on the mail-order catalogue, not entirely a new idea, but one that fitted very well in the American environment.

Such catalogues were put to work vigorously by Montgomery Ward around 1872, and some years later by W.R. Sears (later Sears–Roebuck). They wanted to make every sort of non-perishable goods available anywhere, using the catalogue delivered by mail for the ordering and then delivering by train or wagon. It was an innovative way to banish the 'tyranny of distance'; farmers could enjoy whatever the city folk enjoyed, if they could afford it.

The Beginnings of Electronics

Thomas Edison, John Ambrose Fleming

American inventor Thomas Edison has been called the 'man who created the future', but among his many discoveries was one he did not know he was making. The 'Edison Effect', accidentally found in 1882, involves his light bulb (p. 137). He was trying to stop carbon particles coming off the glowing filament and blackening the inside of the glass.

Edison's lamps used direct current (p. 151); one end of the filament was therefore positively charged, the other negatively. He noted that the carbon particles seemed to come mostly from the negative end. He wondered if there was a way to repel them before they hit the glass, and had light bulbs built with an extra metal plate inside. When the plate was negatively charged, the carbon particles were repelled back to the filament. When it was positively charged, he found a tiny but steady current flowing out of the plate, apparently coming from the filament through the near vacuum inside the bulb. When the plate was negative, no current flowed. Though he patented his effect and discussed it with colleagues, he did nothing more as he was busy with more pressing issues.

The current was actually minute particles soon to be called 'electrons'; the technology that grew from the Edison Effect would be dubbed 'electronics'. Pioneering English 'electrical engineer' (a new term in his day) John Ambrose Fleming kick-started that in 1904. The new technology of 'wireless' (p. 160) had posed some challenges, particularly how best to 'rectify' or 'detect' the incoming signal, that is, make the alternating current flow only one way so that the receiver could make sense of it. Many ideas had been tried, including the 'cat's whisker', a piece of thin wire resting on a crystal of the mineral galena. It was crude but it worked.

Fleming, a consultant to the Marconi company, saw a possible answer in the Edison Effect. A glass bulb containing a glowing filament and a metal plate would allow current through in only one direction (from the filament to the plate), so it was like one of the valves in a blood vessel. The device was soon called a 'wireless valve'. Such valves, given vital extra capabilities by Lee de Forest (p. 176), would remain in use for half a century until the arrival of the transistor (p. 228), in the meantime making possible the growth of radio and television. Edison's chance discovery gave birth to a massive industry.

All the Fun
of the Fair

George Ferris, John Miller

Visitors to the 1893 World Fair in Chicago, celebrating 400 years since Columbus, marvelled at a cascade of technological wonders, including electric lights courtesy of industry pioneer George Westinghouse (p. 129).

Another George exhibited something literally revolutionary. George Ferris, who ran a steel-testing firm, is supposed to have invented the Ferris Wheel while dining with fellow engineers. He drew a sketch and scribbled the specification on a table napkin and the basic design was never changed.

Whatever the truth of that, the Ferris Wheel, designed purely for exciting entertainment, was a wonder, rivalling (as it was intended to), the impact of the Eiffel Tower erected in Paris four years earlier. Hundreds of passengers were lifted simultaneously in wooden cars to a height of 80 metres. Turned by two 1000 horse-power engines, the 15-metre, 70-tonne main axle was the largest single piece of steel forged to that time. The original wheel was destroyed by fire in 1906, but its descendants enliven amusement parks across the planet, and include the immense London Eye on the banks of the Thames.

Gaining prominence at much the same time was the roller coaster or 'big dipper'.

The idea of a 'switch-back', with carts pulled up a slope and allowed to run down again, went back several decades in both France and the USA, before L.A. Thomson built one at New York's Coney Island in 1884. An early dipper simply went up and down a series of hills (dips) in a straight line but it was, no doubt, exciting enough in its day.

Prolific American inventor John Miller, holder of 100 patents, brought us the possibility of the modern roller coaster. His key innovation, patented in 1912 and still used today, ensures a passenger car remains firmly attached to the track, however it moves. Three sets of wheels are involved; one set above the track carries the weight of the car, one inside the track guides the car around turns, one underneath the track locks the car to it.

Miller was initially aiming for greater safety, so that the car would not jump the track, even as the 'dips' were made deeper and steeper. He soon realised more was possible; by the 1920s he was building roller coasters with rapid twists and turns, increasing excitement and popularity from the heightened sense of danger and forming the basis for today's breath-taking rides.

1884

Spinning Turbines with Steam

Charles Parsons

The word 'turbine' comes from the Latin for 'something that spins', but nowadays it means a machine that looks a little like a household fan. Turbines have a long history and many contributors, though some of these merely dreamed and designed, rather than built. In 1500, Leonardo da Vinci (p. 12) sketched one powered by air rising up a chimney above a fire. Fan-like blades, spinning in the air stream, turned meat roasting on a spit. We don't know if he ever built one. A century later, in 1629, another Italian, Giovanni Branca, drew plans for a boiler directing steam onto a turbine like a horizontal water wheel. This was to turn machinery, notably a 'stamping mill', to crush rocks. Again we are not sure if this was more than an idea.

Real progress took a long time. The key issue was efficiency, getting the most value from the burning coal. To absorb an acceptable fraction of the energy in the steam, the turbine had to spin very fast, many thousands of times a minute. Until the late nineteenth century no metal could withstand both those stresses and the heat of the high-pressure steam. Many good minds were onto the problem, including ingenious Dutchman Gustaf de Laval, inventor of the cream separator (p. 136).

Irish engineer Charles Parsons was the son of the Earl of Rosse, builder of the great telescope known as the 'Leviathan of Parsonstown' (p. 64). In 1884 he designed a turbine with a series of angled blades one behind the other on the one shaft. Steam bounced from one blade to another as it passed through, delivering energy at each blow, the blades getting larger as the steam pressure fell. Turning 300 times a second, the first such turbine generated 10 horsepower. It was a brilliant concept, making the turbine efficient enough to compete with, and later surpass, steam engines with pistons.

Parsons' early interest was in ships. A demonstration craft, the *Turbina,* zipped through the water in 1894 at an unheard-of 60 knots. Over coming decades, steam turbines went into many naval vessels and liners, including famous Cunard ships like the *Lusitania,* replacing conventional steam engines until themselves replaced later by diesel engines and gas turbines.

Parsons initially planned the turbine to make electricity, and designed a compatible high-speed generator for the task. Today, virtually all coal-fired power stations use steam turbines descended from Parsons' first models, though improvements in design mean they can turn rather more slowly. Turbines are a vital element of today's jet engines (p. 208).

New Weapons Against Disease

Louis Pasteur, Robert Koch

In the story of vaccines, discovery and invention go side by side. Vaccination against diseases was not possible until we knew what caused disease—but discovery was not enough. We had to find ways to exploit it, so conferring protection through immunity. The quest involved two of the greatest names in medical research, Louis Pasteur and Robert Koch. By the time German doctor Robert Koch arrived on the scene around 1870, his illustrious French contemporary Louis Pasteur had won the early battles on behalf of the 'germ theory' of disease. It was increasingly accepted that diseases in humans, animals and plants are due not to 'miasmas' of infected air, but to infectious microorganisms, or 'bacteria'.

Koch won his reputation through the nasty disease anthrax, suffered by animals and capable of spreading to humans. At the time, he was a busy medical officer in a rural area but still found time for research using a microscope his wife had given him. Inspired by Pasteur, others had found a rod-shaped microorganism (a 'bacillus') living in the blood of infected animals. Samples of blood from animals that had died from anthrax were placed in cuts in the skins of mice. This blood contained the suspect microbe. The mice quickly died of anthrax. Mice inoculated with blood from healthy animals (blood free from the anthrax bacillus) did not die.

A hard man to convince, Koch grew a 'culture' of the bacillus through several generations as free as possible from other contamination. Even though the microorganisms had never directly contacted a diseased animal, they could still spread the disease. The case appeared proven, but it was not until Pasteur developed an anthrax vaccine in 1882 that Koch's findings were really accepted.

Koch added to his achievements by identifying beyond doubt the microbes that caused other diseases, including tuberculosis, cholera and septicaemia. Tuberculosis, known as 'consumption', killed one in every seven people in Europe at the time. Koch's work was therefore judged so important he won the 1905 Nobel Prize for Medicine.

Koch developed ingenious and powerful methods to separate and culture various sorts of bacteria. Armed with his methods, Koch's students identified the germs that cause diphtheria, typhoid, bubonic plague, tetanus, syphilis and pneumonia. So his influence outlived his death in 1910.

The eminent French biologist Louis Pasteur was one of the first to react in 1876 when German Robert Koch found the microorganism that caused anthrax. He quickly corroborated Koch but also noted that not all animals catch anthrax.

Chickens are immune. Was this because their body temperature is around 43°C rather than the 37°C in other animals? Perhaps heat weakened or even killed the anthrax germs. Pasteur chilled the body of a living chicken to 37°C; it lost its immunity.

In a publicly staged trial to answer his critics, Pasteur heated anthrax bacilli and then injected them into 25 sheep. Nothing obvious happened. He then injected the same sheep and 25 others with fresh anthrax. The first lot lived, the second lot died. The heated bacteria had been unable to cause the disease, but somehow made the animals immune. Soon millions of sheep and cattle were being protected against anthrax. Someone estimated that the savings were equivalent to the reparations France had to pay after losing the war against Prussia in 1871.

Pasteur repeated the methods successfully with chicken cholera, another costly animal disease. He called the treatment 'vaccination', so honouring Edward Jenner who 90 years before had pioneered something similar, using cowpox to protect people against the more virulent smallpox (p. 72). He called the cultures of weakened bacteria 'vaccines'.

Pasteur's greatest triumph was over rabies or 'hydrophobia', which could kill humans bitten by an infected dog. He made a vaccine from the spinal cord of rabbits with the disease, though he had not yet clearly identified the bacteria. He could soon stop the disease from spreading from dog to dog but would not try the treatment on a human for fear of something going wrong. He took the step only in 1886, when nine-year-old Joseph Meister was brought to his laboratory after being mauled by a rabid dog.

The boy seemed certain to die, so Pasteur treated him (though he was not licensed to practise medicine). Meister made a complete recovery, as did a young shepherd similarly bitten a few months later. Publicity was intense; Pasteur became perhaps the most famous scientist in the world. Public gratitude supported the founding of the Pasteur Institute, where Joseph Miester was to serve as gatekeeper for many years. Pasteur died from the complications of a stroke a decade later, aged 75.

More than 100 years later, most human and animal diseases, whether due to bacteria or the more recently discovered viruses, can be controlled by vaccination, and there have been some remarkable public health triumphs, including the total elimination of smallpox globally and almost the same success with polio. Deaths from regularly recurring new strains of influenza are routinely reduced through production of new vaccines. In a few diseases, most notably HIV/AIDS, a vaccine still eludes medical science.

The High-voltage Wizard

Nikola Tesla

Hundreds of people attended the funeral of Croatian-born Nikola Tesla in New York in 1943, mourning the loss of an inventive genius. The 87-year-old Tesla held 700 patents, but died an eccentric recluse, a misanthrope with few friends (humorist Mark Twain was one), a poetic visionary with a practical touch.

Arriving in America in 1884 almost penniless, Tesla worked briefly for Thomas Edison. The pair fell out over Edison's excessively methodical methods. Tesla later commented, 'a little theory and calculation would have saved him 90 per cent of his labour'. They also disagreed fundamentally over electricity, with Edison committed to direct current (DC) and Tesla to alternating current (AC). The burgeoning enterprise to distribute electricity for sale pitted Edison against George Westinghouse, who endorsed Tesla's ideas (p. 151).

Tesla knew well the merits of AC: generation was simpler, and AC could be 'transformed' to higher voltages (p. 95), reducing transmission losses over long distances. AC had one drawback. It lacked a motor to convert it back into movement. Tesla supplied that by inventing the 'induction' motor which ran on AC current. He made possible Westinghouse's electrical triumphs, lighting the 1893 Columbian Exhibition in Chicago with AC power, and two years later bringing light to the city of Buffalo from Niagara Falls.

Lacking business skills, and always short of funds for research, Tesla became a master of high voltages, generating artificial lightning, illuminating lamps without wires and staging spectacular shows in his laboratory, where he passed high-voltage AC over his body to demonstrate its safety. Today, his influence is pervasive. Every petrol-driven car uses a Tesla coil—a modified transformer—to generate the high voltages needed to make the spark. Cathode-ray tubes in TV sets rely on similar technology.

Tesla was several years ahead of Marconi (p. 160) in devising ways to transmit and receive 'radio' waves. Marconi secured the patents (including US ones) and enduring fame; but Tesla kept going, building a massive tower on Long Island in 1900 to transmit messages, weather and stock-market reports, and even pictures, worldwide.

In Tesla's one great failure, his proposal to flood electric power into the air, lighting globes at great distances, collapsed when his principal backer, banker J.P. Morgan, withdrew. The dispute with Marconi over precedence in American patents was later reviewed by the US Supreme Court, which decided in Tesla's favour, two years after his death.

The 'Horseless Carriage'

Gottlieb Daimler, Karl Benz and Others

Among the people employed by Nicholaus Otto, inventor of the gasoline-powered four-stroke engine, were Gottlieb Daimler and Wilhelm Maybach; they pushed the technology rapidly forward. Their 1885 engine was lightweight and fast. A year later Daimler installed one in an existing stagecoach to make the first four-wheeled automobile. In 1889 the pair built a complete automobile from the ground up, with a two-cylinder engine and a top speed of 10 kilometres an hour.

By 1890 Daimler was designing and building cars. In 1894 the winner of the first-ever car race had a Daimler engine. Maybach parted company with Daimler soon after, going on to design the Mercedes (named after his wife) and later engines for the new Zeppelin airships (p. 165).

Leading the way was Daimler's rival, fellow German Karl Benz, an iron foundry owner from Mannheim. Seeking additional income, he marketed stationary engines, and in 1885, he married a scaled-down engine to a two-seater tricycle, producing the first gasoline-powered automobile. This had electric ignition, mechanically-operated engine valves, a water-cooled engine and differential gears, all found in cars more than a century later. Benz sold some, but later moved to the more stable four-wheeled format. In 1893 his Ideal model became the world's first mass-produced car; 150 were made in 1895. A 3-horsepower motor allowed a top speed of 14 kilometres per hour, but it struggled on hills.

Supported by the business acumen and unflagging optimism of his wife Bertha, Benz was the biggest maker of cars in the world by 1900. However, he stuck to old designs and paid no attention to increasing speed. His vehicles were soon being overtaken (in both senses) by Daimler's more innovative designs. The Daimler and Benz companies merged in 1926.

Gasoline-powered cars began to outnumber steam vehicles, not that there were very many of either, given the cost. Frenchmen René Panhard and Emile Levassor set the pace with new designs, moving the engine and radiator to the front of the car, with power transmitted to the rear wheels to improve balance and steering.

The making of cars soon reached the US. By 1890, the Duryea brothers were selling 'motor wagons' in Massachusetts. The Duryea was an expensive limousine, in production until the 1920s. America's first mass-produced car was the Oldsmobile, manufactured in 1901 by Ransom Olds in Detroit, Michigan, soon the hub of the US motor industry.

Abundant Aluminium

Paul Héroult, Charles Hall

Aluminium (aluminum in the USA) was never invented; it has been abundant in rocks, soil and gemstones since the Earth formed. From the late eighteenth century, chemists had suspected that such an element existed in the mineral alum, already in use in papermaking and textile dyeing, and in medicine; hence the name aluminium. Isolating it was hard. Decades later they produced small impure samples, but the costly process made aluminium a semiprecious metal. Extracting it in larger, purer quantities for industrial use required major innovation, and depended on cheap and abundant electricity, available only in the late nineteenth century.

Aluminium is so tightly bound to oxygen in the plentiful compound alumina that heating and chemical attack will not release it. The invention of electrolysis by Humphry Davy (p. 78) provided a way of prising out the aluminium—a process also used on other active metals, such as sodium, calcium and magnesium. An economical process did not arrive until 1886, courtesy of Charles Hall in the USA and Paul Héroult in France, working independently and both aged only 22 at the time. Starting with the common ore bauxite, they firstly cleaned away impurities such as iron and silica, then dissolved the purified alumina in molten cryolite, a compound containing fluorine. The melt was placed in large carbon-lined vats. Heavy doses of electrical current were passed through; liberated molten aluminium collected at the bottom of the vats ready to be tapped off. With improvements, this is still the method of production today.

With a unique combination of strength, lightness and resistance to corrosion, aluminium was soon in widespread use, rivalling steel in many applications. The famous statue of Eros in London's Piccadilly Circus was cast in aluminium in 1893. Aluminium's properties can be enhanced by alloying it with other elements. Duralumin, invented accidentally in 1909, contains copper and magnesium as well as aluminium. As strong as mild steel but weighing only one-third as much, this made possible the first airships (p. 165) and radically transformed the early aircraft industry (p. 172). During World War II, due to the high military demand for aluminium, housewives were urged to donate their aluminium pots to the war effort.

We now find aluminium everywhere: in food containers and foil wrappings; electrical conductors; motor vehicle engines and fittings; furniture; and lightweight girders. The high demand its production places on electricity supplies poses a challenge for the future.

Contact Lenses Improve Sight

Adolf Fick, Kevin Touhy, Otto Wichterle and Others

The concept of the contact lens is old. Leonardo da Vinci sketched one in the sixteenth century, Rene Descartes in the seventeenth. Both envisaged lenses small and light enough to be worn directly on the eyeball, rather than in a frame like the familiar spectacles, but no way then existed to make them.

In 1887 the first attempt was made to produce a practical contact lens, though English astronomer John Herschel had already proposed using a mould to ensure a lens fitted snugly against the eyeball. That year German doctor Adolf Fick made contact lenses from glass to correct both long sight and short sight, trialling them first in animals. The lenses, which covered the whole eyeball, were heavy; they severely irritated the eye, limiting use to a few hours.

Two years later German glassblower August Müller made lighter and thinner lenses, more comfortable but still limited in use. Eugene Kalt was another active in the field. As was later discovered, the cornea, the clear tissue covering the front of the eyeball, has no blood supply and must draw oxygen directly from the air. Glass contact lenses did not let oxygen pass, and tissues could be permanently damaged if the lens were worn too long.

A major step forward was not made until the mid-twentieth century, stimulated by the invention of a transparent plastic called PMMA (polymethylmethacrylate). American optometrist William Feinbloom was first to experiment, mixing glass and PMMA and making bifocal, and even trifocal contacts. A decade later, lenses of pure PMMA came onto the market, devised initially by American Kevin Touhy and remaining dominant for 40 years. These were more comfortable and could be worn all day, though oxygen could not seep through.

Even better was HEMA (hydroxyethy-methacryate), invented in the 1950s, not only light and durable but soft. It took a decade for Czech chemist Otto Wichterle to find how to mould lenses from HEMA, but in 1972 German firm Bosch and Lomb began marketing his 'hydrogel' lenses. Permeable to oxygen, and causing very little irritation, they could be worn continually if desired without risking cornea damage. These have now almost totally replaced hard plastic lenses and were soon available in tinted plastic, letting people change their eye colour. Held in place by air pressure against the film of tears between the lens and the cornea, these contacts are almost impervious to shock and can even be worn during contact sports.

Inventing a Language

Ludwik Zamenhof

Nearly all the languages spoken on this planet throughout history have evolved over time as a consequence of population movements, invasions and cultural influences. English, for instance, combines words and usages from Celtic, Latin, Anglo-Saxon, Danish, French and ancient Greek. But some have been invented by someone identifiable, such as the elven languages used in J.R.R. Tolkien's Middle Earth, and the Klingon speech from 'Star Trek'.

Perhaps thousands of 'constructed' or 'planned' languages can be identified, mostly now forgotten. Surviving examples include the several versions of 'Pidgin', as well as Bahasa Indonesia, the official language of Indonesia, created from Malay dialects by a Dutch linguist in the 1920s as a way of unifying the region's diverse cultural and ethnic groups.

The push for unity and reduction of conflict was also the motivation for the best known of the non-official 'artificial' languages, Esperanto, devised by eye specialist and linguist Ludwik Zamenhof, a resident of Poland in the late nineteenth century. From his experiences in a polyglot society, Zamenhof came to believe fervently that a common, easy-to-learn language would help resolve tensions between countries and within communities. The name means 'one who is hoping' in the language, and was originally used as a pseudonym by Zamenhof, only later becoming the name of the language

He began to construct Esperanto when still at school, publishing his first textbook in 1887 when he was 38. The language is deliberately simple in construction and use, with strictly phonetic spelling, grammar that can be written on a single page and a small number of 'roots' drawn mostly from southern European languages, with some of German and Slavic origin. Suffixes and prefixes provide the complexity necessary for conversation.

Despite the efforts of its adherents, Esperanto has no official status, nationally or internationally, and has been actively opposed at times, including in Tsarist and Soviet Russia, wartime Japan, Nazi Germany and Franco's Spain. It has, however, many adherents today, numbering around two million, with many in eastern Europe and Asia. Current Esperanto speakers probably outnumber all speakers of other artificial languages over the last century by ten to one.

Many would argue that the 'universal language' role Zamenhof sought for Esperanto is now occupied by English, given its use in the USA and as the common language of science and technology. Esperantists would not agree.

The Straw and the Flask

Marvin Stone, James Dewar

How good to suck an ice-cold drink through a straw, or to enjoy a cup of tea from a flask that kept it hot. These are everyday pleasures now, but someone had to make the relevant inventions.

The Drinking Straw

Before 1888 the only straws for sucking up your favourite drink were natural ones—stems of grasses like rye. Legend says that American Marvin Stone, whose factory made paper cigarette holders, was enjoying a mint julep after work. He found, as was not uncommon, that the rye stem disintegrated, leaving a gritty residue in the drink.

Thinking, as all inventors do, that there had to be a better way, he tried wrapping strips of paper around a pencil. Removing the pencil and applying a little glue, he had a prototype paper drinking straw. Simple, but profound. After some successful testing, and moving to wax-coated paper to prevent straws becoming soggy, he was soon making more straws than cigarette holders. By 1906 straws were being wound by machine, and spiral-wound tubing was finding a multitude of uses, such as in cardboard mailing tubes and insulation around electrical wiring.

The 'Thermos'

Making sure the drink you are pouring out is cold (or hot as the need might be) required some innovation too. The 'vacuum flask 'was originally strictly for scientific use, devised by leading British researcher James Dewar around 1892 to keep liquefied gases cold, and dubbed the Dewar flask. The secret was two glass flasks one inside the other, with silvered walls and a vacuum between to slow heat getting in or out.

In 1904 two German glass-blowers founded a firm called Thermos to make vacuum flasks for domestic and commercial use. They used the Greek word for 'heat', at the suggestion of a man from Munich who won a competition to devise a name. Since then, many millions have been made, and almost everything needing to be kept hot or cold, or even lukewarm, has travelled in one. Explorers like Shackleton and Perry carried them in polar regions. World War II bombers over Europe carried a dozen each to supply hot drinks to the crew. They have kept vaccines safely cold in the tropics and scientific samples at an even temperature. The popular name persists, though there are now many manufacturers. A generic name is 'vacuum flask'.

AC vs DC

Thomas Edison, George Westinghouse

American inventor Thomas Edison was first to develop a complete system to generate and distribute electricity (p. 137), but he was not unchallenged. George Westinghouse (p. 129), inventor of the air-brake among other things, was also in the game. He lit the city of Buffalo in 1886.

The two had rival technologies. Edison used direct current (DC), which moved only in one direction along the wire. Westinghouse was exploiting alternating current (AC) in which the current reversed its direction many times a second. As Westinghouse saw it, AC had distinct advantages. A 'transformer' (p. 95) could raise the voltage, lowering the current and reducing losses in transmission over long distances. DC could not be transformed, and voltages low enough to be safe needed thick copper wires for acceptable losses. The cost of copper was rising; Edison feared he could not afford to run cables to customers more than a kilometre or two away.

To scare customers away from Westinghouse, Edison claimed that AC was unsafe. In a bizarre act in 1887, he set up a 1000-volt Westinghouse AC generator and executed a dozen animals on an electrified metal plate in front of the press and the public. The term 'electrocution' was coined.

The New York Legislature was then considering new ways to execute criminals, hanging being thought too slow and painful. By 1888, they had decided on electrocution in an 'electric chair', with both AC and DC as possible power sources. Edison campaigned for the AC version: surely customers would not buy current from a man who also powered the electric chair. The committee deciding which system to use was headed by Fred Peterson, who was on the Edison payroll. The committee chose AC.

Westinghouse refused to supply the needed generators. He also funded the appeal made on behalf of the first man sentenced to be so executed, on the grounds that the Constitution prohibited 'cruel and unnatural punishment'. The appeal failed; the first man to sit in the electric chair died from AC current supplied by an Edison-made generator.

The ploy did Edison no good in the long run. After powering the 1893 Exposition in Chicago, Westinghouse brought power from the first hydroelectric station in Niagara Falls in 1895, clearly demonstrating the benefits of AC in hauling electricity over long distances. By this time, Edison was changing over and AC has remained the basis of electricity distribution ever since.

The Advance of Synthetics

Hilaire de Chardonnet, Jacques Brandenburg, Leo Baekeland

What we now call 'synthetic materials' or 'plastics' started in the mid-nineteenth century with celluloid (p. 118), which soon featured in billiard balls, shirt collars and photographic film. It was not fully synthetic; its starting point was the natural material cellulose, found in cotton and other plant fibres, reacted with nitric acid and other chemicals.

Chemists continued to tinker with cellulose in search of other products. A new approach was to break it down into tiny particles by various chemical treatments and then make those identical fragments link up again (to 'polymerise'). The broken-down cellulose was called 'viscose'.

There were two ways to proceed. If the viscose was forced through many small holes and into an acid bath, the fragments formed into fine fibres, able to be drawn into a thread for spinning and weaving. Dubbed 'rayon' from its sheen, or 'artificial silk', and first produced by Frenchman Hilaire de Chardonnet in 1889, it was soon very popular (and remains so).

Alternatively the viscose was sent through a narrow slit into the bath, forming a sheet of 'cellophane'. Swiss chemist Jacques Brandenburg first achieved this in 1912, after a long search for a flexible, water-resistant film to protect food and other perishables. The US company DuPont began large-scale production under license after World War I. Very popular and useful in its day, cellophane is still about, but for many purposes has been replaced by newer and better materials.

Synthetic materials more removed from the natural world began to emerge in the early twentieth century. Best known of these was bakelite, invented in 1904 by Leo Baekeland, Belgian-born but living in USA. He mixed formaldehyde with phenol, found in coal tar, and made a 'resin'. When precisely heated and pressed, the resin hardened, taking up the shape of its container. Thereafter it would not burn, melt or dissolve in corrosive chemicals; nor would it crack, shatter, fade or discolour when exposed to temperature changes, sunlight, dampness or sea salt. Bakelite was unlike any material known.

This first 'thermosetting' plastic (hardening when heated) found a great many applications. Its high electrical resistance found use as an insulator in electrical circuits or in the new 'wireless valves' (p. 176). The military found its combined lightness and strength better than steel for many purposes and almost all World War II weapons contained some bakelite. It is obsolete now.

The Power in the Punch Card

Herman Hollerith

Now little remembered, Herman Hollerith has several claims as one of the progenitors of today's information technology. For one thing, a company that he founded late in the nineteenth century became, after several mergers and changes of name, International Business Machines or IBM, highly visible in the computer market a century later. And he automated the counting of the 1890 US census, the biggest data-processing endeavour ever undertaken.

Hollerith was only in his twenties, and on the staff at the Massachusetts Institute of technology, when he devised a process to encode data (such as the census collectors gathered) as patterns of holes on cards. These were not unlike the cards used to control a Jacquard loom (p. 80), but he was reputedly more influenced by the way tram and bus conductors punched holes in tickets to indicate the fare paid. Stacks of cards were sent through reading devices with spring-loaded pins that completed electrical circuits when they passed through the holes. Numbers corresponding to particular categories were automatically totalled on dials. The cards could also be sorted to select and count people who corresponded to a particular grouping of characteristics, such as white women living in the Bronx, aged between 20 and 30 and employed as shop assistants.

In trials, Hollerith's method easily out paced the existing manual counting. The government put his machines to work in time for the 1890 census. It did the whole job in seven years instead of nine, even though the census was far more comprehensive than previous head counts and called for more complex analysis. But it cost twice as much as the census a decade earlier, largely due to the immense amounts of power consumed by the electrically-driven card readers and sorters.

Hollerith's triumph was widely noted. *Scientific American* devoted its August 1890 edition cover to the innovative technology. Hollerith's Tabulating Machine Company struggled to keep up with demand. Orders for the machines (rented rather than sold), came from other census authorities around the world and from government departments and private firms swamped with paperwork and mountains of figures. Some companies such as Prudential Insurance devised their own punch-card systems.

None of these were computers as we know them today but they did foreshadow the powerful impact automatic data processing would make. Punching and sorting cards remained a activity in offices (for 'key-punch operators') for most of the next century.

Done and Undone with Speed

Whitcomb Judson, Gideon Sunback, Georges de Mestral

For centuries ancient techniques have held our clothes together and on our bodies: buttons and buttonholes, hooks and eyes, and laces. Chicago engineer Whitcomb Judson tried something different.

In 1893 Judson patented a 'slide fastener'. Two rows of metal teeth faced each other and were made to interlock and then unfasten again by moving a slide one way or the other.

The cunning mechanism is little changed today. Moving the slide slightly widens the space between adjacent teeth so opposite teeth can mesh as they are brought into contact. The initial market was for high boots; fastening those took time. But the device sold slowly. The crude initial models were likely to jam or come undone.

Swedish designer Gideon Sunback made the operation smoother and more reliable. Orders came for fasteners for army uniforms during World War I, but the clothing trade as a whole was uncertain. It appears the great clothes designer Schiaparelli made the new fasteners acceptable by incorporating them into garments from 1930, giving apparent approval to a device that let someone exit their clothes quickly. It was now called the 'zip-fastener' or the 'zipper', named by a PR man or a writer summing up its smooth speed as 'zip'.

We don't know where Judson got his basic idea from, but for the zipper's more recent rival, the story is well known. Fifty years after Judson's innovations, Swiss engineer Georges de Mestral regularly returned from walking his dog in the Alps with his dog's fur and his own clothes covered in burdock seeds. Under a microscope, he saw how the seed's shape let them hook onto clothing fibres and dog hair. He determined to emulate it, though others laughed.

In 'hook and loop' fastening that Mestral invented into 1948, one side of the join is covered with thousands of tiny plastic hooks, the other with even smaller loops. The hooks grab the loops, and a large force is needed to break so many linkages over a wide front. The join is always much longer than it is wide; pulling the sides apart with a sideways peeling action requires only a small number of links to be undone at a time.

The product is now called velcro, from French words for 'velvet' and 'hook'. Like the zipper, it has drawbacks, accumulating lint and dirt, attaching to clothing. Yet it has proved very versatile and is widely used in footwear and clothing, replacing buttons, zippers and laces. It has even found a place in surgery.

Diesel: the Man and the Engine

Rudolf Diesel

Rudolf Diesel remains well-known today through the type of engine that bears his name, now widely used, and its distinctive fuel. Diesel should have made a fortune from his inventions, but had no head for business. He disappeared from a Channel steamer in 1915, most likely falling overboard, though some suspect he was driven to suicide by financial worries.

Diesel trained as a refrigeration engineer in Munich; his exam results were the highest ever achieved. He had a mixture of motives in seeking new sources of power. Steam engines, though much improved by James Watt and others, remained unacceptably inefficient, turning perhaps 15 per cent of the energy in coal into useful work. They were also too big and costly for small factories and workshops. More efficient (therefore smaller) engines could power vehicles.

Etienne Lenoir and Nicolaus Otto had already created 'internal combustion engines' in which the fuel was burnt inside the cylinder rather than outside, as in a steam engine. Seeking greater efficiency in a 'rational heat engine', Diesel proposed injecting fuel into the cylinder and igniting it, not with a spark or flame, but by the high temperature generated as the piston compressed the air in the cylinder. By working through a much wider range of temperatures and pressures, his engines would exploit more of the energy of the burning fuel. He took 14 years to turn these ideas, much disputed at the time, into a patent.

Diesel's first engine, set running in 1893, was a failure. The intended fuel was coal dust, piled up in vast heaps around the German Ruhr as waste; funding came from industry giants like steel magnate Baron von Krupp. The coal dust-fuelled engine exploded during trials, nearly killing him.

A second model, also operational in 1893, used heavy fuel oil, less refined and therefore cheaper than gasoline. His 1895 model had unprecedented fuel efficiency, opening up many applications. Diesel became famous. Within a few decades, diesel engines were powering factories, mills, water pumps, automobiles, trucks and boats, trains and ships.

Diesel engines have traditionally produced smoky exhaust fumes, posing risks to health and the environment, and poor acceleration, making them less popular for automobiles than petrol-driven engines. But more refined technology, especially in fuel injection, and better quality fuels have largely solved these problems, letting the inherent benefits of Rudolph Diesel's creation shine through.

X-rays: the Unexpected Radiation

Wilhelm Röntgen

In the late nineteenth century, German Wilhelm Röntgen, like many other physicists, was interested in the newly discovered 'cathode rays'. These were generated when high-voltage electricity was passed between two electrodes inside a glass tube almost empty of air. It would soon be shown that cathode rays were streams of tiny fragments of electric charge to be called 'electrons', the same particles that had caused the Edison Effect (p. 140). The phenomenon would give rise in due course to television (p. 161) and other uses of the 'cathode ray tube'.

Various researchers had noted, to their annoyance, that photographic plates stored near such 'discharge tubes' were inexplicably fogged, as if they had been exposed to light. Röntgen wondered if discharge tubes gave off invisible radiation, especially since the walls of the tubes glowed under the impact of the cathode rays. Perhaps the radiation could leak through the packaging surrounding photographic plates and expose them.

One November evening in 1895, he found this was so. His suspected radiation easily passed through a light-proof shroud around the discharge tube, and made a plate covered with a fluorescent chemical glow several metres away. For something so unexpected and mysterious, the term 'X-rays' seemed appropriate. They were also called Röntgen Rays for a time.

X-rays were stopped by a thin sheet of metal, but could cut through human tissue and emerge with enough energy to leave an image on a photographic plate. The first 'X-ray photograph' ever taken was of Röntgen's wife's hand, clearly showing her bones and a ring she was wearing, surrounded by the shadow of her flesh.

Röntgen became, justly, one of the most famous scientists of his age, winning the first Nobel Prize for Physics ever awarded. He remained strikingly reticent and modest, preferring to work alone, building his equipment with his own hands. A lover of nature and a daring mountaineer, he was finally taken by bowel cancer at the age of 78.

Röntgen cannot be said to have 'invented' X-rays; they had always existed. But he did show how they might be used. X-ray images have proved of incomparable value to medicine and surgery (even every up-to-date dentist has a machine) and later to engineering and industry. More powerful and precise X-ray tubes were soon invented, particularly by the American William Coolidge from General Electric. The hazards of the technology, such as its capacity to cause cancer, were discovered only later.

Machines that Make Music

The Pianola and the Juke Box

Not everyone can create music for themselves. Hence the wish for machines producing music at the turn of handle or the press of a button. In the long-loved musicbox, the tune is coded as an arrangement of spikes on a drum. As the drum turns, the spikes touch metal strips of varying lengths to create the different notes.

The player piano was a step up in complexity and sophistication, drawing music from the familiar piano (p. 40) without human aid (other than to pump the pedals and drive the works) The first such, the 'pianola', was marketed by the Aeolian Company of New York in 1895, the creation of Edwin Votey. There were soon many manufacturers, including the Tonk brothers, whose pianos had a distinctive sound later dubbed 'honky-tonk'.

To control the performance, a long roll of perforated paper passed through the mechanism; holes indicated when a particular note should be sounded by the aid of air pressure. The rolls were constructed, not by musicians, but by technicians who simply transferred the content of sheet music. 'Pianola' later came to mean almost any type of automatic piano, including the advanced 'reproducing pianos' invented a decade later in Germany. These added nuances of dynamics, rubato and pedalling, and many pianists and composers recorded on them, including Jan Paderewski, George Gershwin and Percy Grainger. Player pianos let music-lovers recreate those performances in their living room, and were a major source of home entertainment until the 1930s, before being superseded by the 'wireless'.

The purpose of the jukebox was similar to the player piano—to produce music from a machine on demand, the difference being that the music was completely pre-recorded. The first of these appeared in 1889, playing Edison cylinder recordings (p. 135). The customer put a nickel in the slot and listened to the 'Nickelodeon' through a tube. Into the new century the demand for both recordings and 'coin in the slot' players boomed. Newer machines could hold more recordings, allowing a greater choice of music. As record players proliferated in the home, jukeboxes concentrated more on dance music, but the rise of radio took its toll here too (p. 189).

The name 'jukebox', from the 1930s, is Afro-American; 'juke' perhaps means 'to dance'. Appropriately, jukeboxes gave black recording artists a chance to be heard; for decades, radio was for white musicians only. Nowadays, music blares all the time at us from video screens, but it is not usually music of our choosing.

The Moving Picture Show

Thomas Edison, the Lumière Brothers

Motion pictures (or 'movies') were arguably the dominant cultural influence in the West in the twentieth century, at least until television arrived, but their origins, a century earlier, lay in toys to provide diversion. These made use of the slow rate of processing of information by the human brain. A series of still images, each slightly different from the next, appear blended together into a moving image if they are presented to the eye at least 10 times a second and with a blank space before each image.

These toys had a variety of names and originated in a number of countries. In the 'thaumatrope' ('wonder turner') of 1826, each side of a card carried a different image, such as a bird and an empty cage. Spinning the card rapidly on a twisted string put the bird (apparently) in the cage. A few years later, the 'phenakistiscope' ('deceptive view') let the viewers see a number of images set around a drum reflected in a mirror as the drum was turned. Abstract patterns moved; acrobats tumbled. The simpler 'zoetrope' ('wheel of life'), initially ignored but revived in the 1860s, allowed more than one person to view at a time, the later 'praxinoscope' ('seeing something happening') set the moving images against a background.

None of these shows could last very long, a few seconds at most, before repeating. Longer performances needed many more images, made possible when George Eastman began marketing photographic film (p. 85) made from the newly invented celluloid (p. 118). 'Magic lantern' shows, which projected photographic images onto a screen, were already popular for entertainment and education. Combining that with the idea behind the persistence-of-vision toys ultimately produced the first movies. However, that took several decades and many active minds, including Thomas Edison's.

A kick-along came from the famous series of images taken by English-born Californian photographer Eadweard Muybridge in 1872. In order to prove that a galloping horse has all its hooves off the ground together at one point during each stride, he arranged dozens of cameras in a line beside the horse's path, exposing them one after the other by tripwires. Edison knew of this work, but thought that a single camera taking multiple images on a long strip of film would be much more practical. By 1891 he had developed the 'kinetoscope' ('seeing motion'), though his assistant William Dickson did most of the work. This was a 'peepshow' for one viewer at a time, marketed like his coin-in-the-slot phonographs (p. 157). These became popular in amusement parlours, showing a minute of moving image for 50 cents.

For others, the future lay in entertaining or amazing a whole audience at once. The leading players in America included young collaborators Charles Jenkins and Thomas Armat. They had devised a motion picture projector (the 'phantosope') by 1895, demonstrating it at an exposition in Atlanta. The pair split up soon after, each claiming to have invented the machine, which indicates the interest it had aroused. Armat sought help from Edison, who agreed to build the machine provided it was marketed as Edison's Vitascope. The first movie screening in the USA took place in a theatre off Herald Square in April 1896. Images of rough seas off the Dover coast were the most popular.

Edison left his imprint on the new technology. His film stock was 35 millimetres wide, still the standard size for professional moviemaking today. He placed holes down the side of the film to pull it through the projector, a revolving shutter to flash the images, a 'Maltese cross' device to stop and start the film many times a second. Emerging moviemakers everywhere used very similar ideas.

Nonetheless, Edison had been beaten to a key milestone by Auguste and Louis Lumière, the 30-something sons of a Paris photographer. The brothers used equipment very similar to Edison's, but combined the camera and projector into one relatively lightweight machine called a cinematograph. They staged the world's first public screening in the basement lounge of a Paris café on 28 November 1895.

The program of 10 films lasted only 20 minutes, but patrons marvelled at moving images of everyday scenes—a rowboat in a harbour, workers leaving a factory, a train pulling into a station (the latter so realistic that people ducked for cover, fearing they were about to be run over).

Though pleased with their achievement, Louis Lumière saw no future in the 'cinema'. Why would people pay to see things they could observe just by stepping outside? Yet the patrons came in increasing numbers to view the new marvel. Into the new century, the technology found its role as entertainment, filming vaudeville acts and scenes from successful plays, and ultimately creating new stories on film, presenting action not possible on the stage. A technological curiosity was about to become a major art form.

The Telegraph Without Wires

Heinrich Hertz, Edouard Branly, Guglielmo Marconi

The saga of 'radio', aka 'wireless' (an early name, now coming back into favour in the digital age) begins in 1863. The Scottish mathematical genius James Maxwell predicted the existence of a then unknown form of radiation, similar to light, generated by rapidly moving electrical charges, such as in a spark. In 1887 physicist Heinrich Hertz (whose name is still used today as the unit of frequency) confirmed that such radiation existed. He found that sparks indeed make such radiation, which behaved as predicted, for example travelling at the speed of light.

Hertz was delighted to have confirmed Maxwell's prediction, at least inside his laboratory, but he was pessimistic that any use would be found for the 'Hertzian Waves'. Some wondered if the waves might be used to transmit messages, say in morse code. But there were practical problems to be solved; one was detecting when the rapidly oscillating radiation reached the receiver.

To meet the need, Frenchman Edouard Branly invented the 'coherer', a glass tube filled with iron filings. Normally this was an insulator, but it would allow electricity to pass when struck by the radiation so creating some audible sign, such as a 'click' in the headphones. This was much better than watching sparks jumping across a gap as Hertz had done, since it effectively amplified the incoming signal. With such primitive technology, Branly succeeded in detecting the radiation 150 metres from the sparks that generated it.

Other bright minds, such as Englishman Oliver Lodge and the young New Zealander Ernest Rutherford, were also at work. The most famous of the early experimenters was Guglielmo Marconi. From 1895 he tried to pick up Hertzian waves at ever-greater distances, starting with a few kilometres within his family's Italian estates (a gun was fired to signal from receiver to transmitter that the signal had come through).

Failing to get any official support, Marconi moved to England, where his English-born mother had relatives in the post office. With their aid, he sent signals from Portsmouth to the Isle of Wight, then across the English Channel, and ultimately in 1901 across the Atlantic from Newfoundland to Cornwall. Stopping and starting the waves in accordance with the morse code (the transatlantic message was the letter 's'), he initiated the age of 'wireless telegraphy'. Marconi later shared the Nobel Prize for his invention. A revolution in communications was stirring.

The Father of the Cathode Ray Tube

Karl Braun

Italian Guglielmo Marconi won the Nobel Prize for Physics in 1909 for pioneering work in wireless telegraphy (p. 160). He shared it with German Karl Braun. During a long career as a university professor, Braun increased the range of Marconi's transmitter and invented the 'crystal rectifier'. In 1874 he had found that certain crystals such as galena (lead suphide) let electric current pass only one way. In early wireless receivers (and generations of 'crystal sets' built by enthusiasts) a coil of thin wire (the 'cat's whisker') was pressed against such a crystal, sending the rectified current (and the message it carried) to the headphones.

All this is long obsolete, replaced by vacuum tubes and then by transistors (p. 228). Today we gratefully recall Braun as the inventor the cathode ray tube (CRT). Just as Marconi had refined Hertz's primitive radio equipment, Braun had taken on the 'discharge tube', a laboratory tool used for decades by physicists, particularly in Germany and England, that lead to the discovery of X-rays (p. 156) and the invention of flourescent lamps (p. 170). In discharge tubes, high-voltage electricity was made to pass between two metal plates (or 'electrodes') in a sealed glass tube almost empty of air.

All sorts of fascinating phenomena were seen. With the right shape and arrangement of electrodes, one end of the glass tube could be made to glow, especially if coated on the inside with a chemical like zinc sulphide. The effect was traced ultimately to the impact of some otherwise invisible 'rays' that came from the negatively charged electrode (or 'cathode'); hence the term cathode rays.

In 1897 Braun honed the old discharge tubes into something much more versatile. Controlling the cathode ray beam with electric and magnetic fields focused the glow into a small spot and allowed it to be scanned from side to side across the flatted end of the tube and in rows from top to bottom. This produced the first 'cathode ray oscilloscope', or CRO. Much improved in operation over the years but essentially unchanged in principle, the CRO remains vital equipment in every electronics workshop and laboratory, displaying varying electric currents.

Stimulated by Scottish-born Alan Campbell-Swindon (p. 180), cathode ray tubes morphed into the first television cameras and monitors, and later still into monitors for radar sets and computers. Only now are CRTs giving way to liquid-crystal displays and plasma screens (p. 266).

Sound from a Tape

Valdemar Poulsen

While Thomas Edison and Emile Berliner were recording sound on cylinders and discs, other experimenters had another strategy; they were using the electric currents that represented sounds (as in the telephone) to alter the magnetic field being imprinted on a steel wire or tape. The Dane Valdemar Poulsen succeeded first in 1898; his magnetic recording of Austrian Emperor Franz Joseph is the oldest known. Poulsen soon turned his attention to radio: dictating machines using wire or tape could not compete with cheaper, more reliable cylinders from Edison and Dictaphone.

Not until the 1930s did practical magnetic recording machines reach the market, especially in Germany. These took advantage of advances in electronics (p. 176) and recorded on wire or tape on reels and in cassettes, and later plastic tape coated with iron oxide, as today. Other innovations improved sound quality. Much of the post-war demand for 'tape recorders' resulted from publicity generated by entertainer Bing Crosby, who needed to record his popular radio shows for later broadcast in other time zones across the USA. Tape had advantages over discs for some purposes. It could be physically cut and joined to allow content to be edited, impossible on discs. Hollywood studios quickly seized on the new technology to record speech, music and sound effects to be edited to form film soundtracks.

Cracking the home market was much harder. Tape recorders were big and heavy, as well as expensive; only music buffs and 'hi-fi' enthusiasts were willing to wrestle with the technology. In 1965 Dutch electronics firm Phillips first marketed the 'compact audio cassette'. A quarter the size of other cassettes, this recorded 30 minutes of music on each side. Cassette recorders/players were equally compact. Phillips originally intended the cassette for business dictation but the market was soon dominated by consumers wanting a cheap, simple way to record and replay music.

The audio cassette era peaked in 1979, 80 years after Poulsen's first recording, when the Japanese company Sony released the Walkman, barely larger than the cassette in it, and able to be carried (or rather worn) by users as they walked. That changed forever the way people listened to music, preparing customers to accept (indeed, desire or demand) the portable CD player (p. 287) and later the MP3 player (p. 291). The latter two innovations have made the cassette player mostly obsolete, of course.

Controlling Pain and Fever

Felix Hoffman

The most widely used medical drug in the world is probably aspirin, known scientifically as acetylsalicylic acid. In an unrefined form, it was found for generations in traditional potions made from the bark of the white willow tree (*Salix alba*, hence the scientific name) and other plants. Aspirin is versatile. It reduces pain and lowers fever, particularly by affecting vital body chemicals called prostaglandins. In more recent times aspirin has been found to thin the blood, reducing the risk of strokes and heart disease. In high doses it gives relief from arthritis.

An early form of the drug, salicylic acid, was extracted from plants at the beginning of the nineteenth century, but its use generated some severe side effects—bleeding in the stomach, even death. Because of this, little interest was shown until 1897, when Felix Hoffman and a colleague at the large German chemical firm Bayer discovered how to make acetylsalicylic acid, a form of the chemical that did not exist in nature and had fewer side effects. The first patient was Hoffman's own father, who suffered severely from arthritis and could not tolerate straight salicylic acid.

Bayer patented the drug in 1899; the modern pharmaceutical industry was born. Although another company was already selling it, and a French chemist had devised a similar form 40 years earlier but thought it impractical for large-scale use, Bayer was first to mass-produce the drug, and it is best known by the Bayer brand name. The impact of the non-addictive aspirin was ultimately global. During World War I various countries devised ways to make the drug synthetically, as Germany had begun to do, from the chemical phenol found in coal tar (p. 114). In Australia, the result was called Aspro.

At the time aspirin came into use, the only other antipyretics (drugs to reduce fever) came from the bark of the cinchona tree, along with the antimalarial quinine. Supplies of the bark ran short in the 1880s and the hunt was soon on for alternatives. This quickly led to the identification of the antipyretic phenacetin and then of paracetamol, though the latter's importance was not recognised for half a century. Not until the 1950s was paracetamol on the market, known as Tylenol in the USA and Panadol in the UK, where it was at first available only on prescription.

Between them, aspirin and paracetamol have fixed a lot of headaches and high temperatures; our lives would have been very different without them.

Sailing
Beneath the Waves
David Bushnell, Robert Fulton, John Holland

The dream of travelling under the sea rather than on it is an old one. In 1573 English mathematician William Bourne proposed an ingenious leather-covered boat able to submerge and surface with the aid of leather buoyancy tanks. The first man reputed to have built such a craft was the Court Inventor to England's James I, Dutchman Cornelius Drebble. According to reports, his craft made several voyages under the Thames around 1620, once with the King onboard, but details are sketchy. We do not know how the boat submerged, how it was steered or how the crew got air.

A submarine first went to war in 1776, following the American Revolution. Looking like a coconut built of oak, the one-man *Turtle,* the brainchild of David Bushnell, was powered by hand-turned screws, raised and lowered by buoyancy tanks, and allowed vision from a 'conning tower'. Its armament comprised a time bomb to be attached by a screw to an enemy warship, but the one attack, on the HMS *Eagle,* was unsuccessful.

Decades later, American steamboat pioneer Robert Fulton designed the *Nautilus* (later the name of Captain Nemo's submarine in *20,000 Leagues Beneath the Sea* and later still of the first nuclear-powered submarine). Like a modern submarine, this had wing-like planes on either side of the hull to direct the path of the craft higher or lower. Napoleon was interested for a while, but it seems the *Nautilus* never went to sea.

In order to progress, submarines needed a way to power them (other than by hand-driven screws) and an efficient weapon for use against ships. Compressed air, steam and gasoline engines and the electric motor provided the first, the torpedo the second. First to pull these elements together was the Irish–American schoolmaster John Holland. His early experiments were paid for by the Irish-expatriate Fenian Society, with the hope of destroying the British Navy. In 1893 Holland beat off many rivals to secure the first contract let by the US Navy for a submarine.

The ultimate outcome of his efforts was the 15-metre *Holland,* the first practical and reliable submarine, purchased by the US Navy in 1900. Though lacking a periscope, it was a modern submarine in most respects, with 'dual propulsion' (gasoline powered on the surface, electric power when submerged), used by all submarines until nuclear power (p. 232) took over 50 years later.

The Rise and Fall of the Airship

Ferdinand von Zeppelin

Pioneering airship builder Count Ferdinand von Zeppelin from Germany was not the first to tackle the main problem with passenger balloons (p. 57); lacking motive power, they were at the mercy of the winds. Another name for airships, 'dirigibles', from the French 'to steer', makes the same point. Airships, unlike balloons, could go mostly where they wanted. Frenchman Henry Giffard had flown 30 kilometres in a steam-powered airship in 1852; over the next 50 years others used petrol engines and electric motors.

Yet Zeppelin's is the name we remember. A Lieutenant-General at 53 before retiring, Zeppelin passionately promoted airships as war machines for reconnaissance and the bombing of cities and ships at sea. He faced major engineering challenges. In larger airships, the streamlined, gas-filled bag that provided lift was more prone to instability, and could collapse disastrously. Zeppelin's answer was to give his airships a rigid framework of girders to support an outer covering and enclose a number of separate gas cells. In later larger models, the framework was made of the new aluminium alloy duralumin (p. 147).

Luftshiff Zeppelin 1 made its maiden voyage in 1900. Over the next decade, 'zeppelins' proliferated in German skies, carrying mail and passengers, and enjoying a remarkable safety record given that the lifting gas was flammable hydrogen. They were accepted into army service in 1908. By 1914 Germany had the largest airship fleet in the world, including 100 ready for military duties. Zeppelin had become a much-decorated hero, his creations a source of national pride.

Yet the vision failed the reality test when war came. While appearing huge and fearsome as they silently approached, zeppelins performed poorly as bombing platforms and did little damage. They were themselves vulnerable to attack from fighter aircraft and ground fire. By war's end, their roles were filled by airplanes (p. 172).

Zeppelin died in 1917, missing the great age of civilian airships. In 1919, the British *R-34* was first across the Atlantic. Though many airships, including Germany's *Graf Zeppelin*, operated successfully for a decade or more, with safer, non-flammable helium gas, several fatal crashes, including the *Hindenburg* in 1937, eroded public confidence. Through the 1930s, airships could not handle the competition for passenger traffic from faster and more reliable aircraft, and were ultimately withdrawn, though airships played a major role in anti-submarine warfare during World War II.

The Car of the Future?

Ferdinand Porsche, Victor Wouk, David Arthur

In 1900 automobile pioneer Ferdinand Porsche exhibited an unusual vehicle at the Paris World Fair. Its gasoline-powered engine did not drive the wheels directly. Instead it turned a generator sending current to electric motors on the wheels. This was the first 'hybrid' car (more precisely, a 'series hybrid'); it was decades ahead of its time. Back then, and for decades to come, petrol was cheap, fuel efficiency unimportant and air pollution not a worry. Only in the 1970s would such issues start to matter, and hybrids again attract attention.

In 1974 American engineer Victor Wouk, brother of novelist Herman Wouk, was much involved with electric cars. He refitted a 1972 Buick Skylark with a much smaller engine connected to an electric motor. The car could be driven by the engine, the motor or both. Wouk had been stimulated by the passing of the *Clean Air Act* in 1970. Such a hybrid, he believed, would cut tailpipe emissions, say of carbon monoxide, by 90 per cent, while also greatly improving the 'miles per gallon'. That proved correct. Testing by the US Environment Protection Agency found the car gave off only 9 per cent of the emissions of a purely gasoline car of the time, and provided double the fuel efficiency of the car before conversion. Wouk, too, was ahead of his time. The program that funded the research was cancelled in 1976, and interest in hybrids was left to hobbyists.

In the 1990s, with rising fuel prices and troublesome air pollution, the US government sought collaboration from industry in developing 'next-generation vehicles', with hybrids much in the frame. By then, we had another vital piece of technology, 'regenerative braking', brainchild of electrical engineer David Arthur. The motor could also serve as a generator. When the car braked, its energy of motion was converted back into electricity and stored in a battery. This improved fuel efficiency still more. Regenerative braking is found in most hybrid vehicles now being marketed (albeit in small numbers so far) by most of the major manufacturers.

Into the twenty-first century, hybrids, with their many advantages, seem certain to loom larger in the market, as costs and performance match those of traditional cars. But there are other options, such as cars powered by fuel cells (p. 252) burning, for example, hydrogen. The demise of the gasoline-only car is still some decades away, but we can see it coming.

1900

The Wonderful Century

Alfred Wallace Sums Up

Alfred Wallace, the great biologist and, with Charles Darwin, the founder of the theory of biological evolution by natural selection, was also an observer of progress in science and technology. Around 1900, he summed up what he saw as the great discoveries and inventions of the previous 100 years in his book *The Wonderful Century*. His assessment is worth noting:

Taking first those inventions and practical applications of science which were perfectly new departures, and which have also so rapidly developed as to have profoundly affected many of our habits, and even our thoughts and our language, we find them to be fourteen in number;

1. **Railways**, which have revolutionised land travel and the distribution of commodities.

2. **Steam navigation**, which has done the same thing for ocean travel and has besides led to the entire reconstruction of the navies of the world.

3. **Electric telegraphs**, which have produced an even greater revolution in the communication of thought.

4. **The telephone**, which transmits, or rather reproduces, the voice of the speaker at a distance;

5. **Friction matches**, which have revolutionised the modes of attaining fire.

6. **Gas lighting**, which has enormously improved outdoor and other illumination;

7. **Electric lighting**, another advance now threatening to supersede gas:

8. **Photography**, an art which used to the external forms of nature what printing is to thought.

9. **The phonograph**, which preserves and reproduces sounds as photography preserves and reproduces forms.

10. **Electric transmission** of power and heat.

11. **The Röntgen rays**, which render many opaque substances transparent and open up a new world to photography and to science.

12. **Spectrum analysis**, which so greatly extends our knowledge of the universe.

13. The use of **anaesthetics**, rendering the most severe surgical operations painless.

14. The use of **antiseptics** in surgical operations which has further extended the means of saving life.

Emphasising the increasing pace of technological change, Wallace would concede only seven inventions of comparable importance in all human history up to 1800, some of these very ancient; writing using an alphabet, Indian/Arabic numerals which made possible modern mathematics, the mariner's compass, printing using movable type, the barometer and thermometer, the telescope and the steam engine.

Making a Good Cup of Coffee

Luigi Bezzera, Sartori Kato, James Mason, Melitta Bentz

Coffee has been popular for centuries. It originated in Ethiopia and spread into the Middle East and beyond. Currently half a trillion cups are drunk around the planet each year.

Discerning coffee-drinkers generally like it made fresh, tailored to their own tastes and produced quickly so they can get on with drinking. Those stringent requirements can be met by an espresso machine, judged by some to be the centrepiece of civilisation.

The big name here was Italian Luigi Bezzera. In 1901, he patented a machine that used steam pressure to force hot water through ground coffee held in metal filters. The coffee was stronger and more quickly made than with earlier low-pressure machines, but results were variable. The operator had to judge the pressure and temperature of the water and the time it ran through the grounds, since those factors all affected strength and flavour. Water hot enough to generate the steam could damage flavour.

Modern machines have largely automated the process, with better ways to develop the needed pressure, such as spring-loaded pistons in 1948 and electric pumps in 1960.

With no espresso machine at hand, a quick alternative is instant coffee.

Japanese-American chemist Satori Kato devised a process in 1901, but sales were minimal until 1938, when the Nestlé company marketed Nescafé, firstly in Switzerland. The water was extracted from concentrated coffee, initially by heating, later by spraying; now freeze-drying is used. Frozen concentrate is exposed to low pressure, causing the ice to evaporate without first turning to water, and leaving the soluble coffee power behind.

Instant coffee was a hit with US soldiers during World War II; there was some in the ration pack of every GI. The government bought up most of the production for several years; civilians had to wait until peace came before they too could freely enjoy instant coffee.

American James Mason and German housewife Melitta Bentz were other coffee innovators. He invented the coffee percolator in 1865; she patented coffee filter paper in 1908. To avoid overbrewing and to get rid of the grounds, she fashioned a conical filter from her children's blotting paper, placed ground coffee in it, and poured in boiling water. The coffee brewed as the water ran through. She and her husband Hugo were soon manufacturing filter papers under the still-familiar Melitta brand.

The Start of the 'Throwaway Society'

King Gillette

Acceptable facial hairiness changed through the nineteenth century, from the rigorously clean-shaven, when Beau Brummel dictated fashion, to the elaborate whiskers of late Victorian times. Some shaving was always needed, increasingly done at home as implements improved.

The typical razor had a tempered steel blade, much like that of a sword; hence the term 'cut-throat razor'. The user could easily hurt himself, and razors were expensive, requiring regular sharpening. In the 1880s the Kampfe Brothers in the USA reduced shaving injuries by fitting a wire grid mesh on the side of the blade that pressed against the skin, so that only the hairs sticking through the guard mesh were cut. However, the blades in this first-ever 'safety razor' remained expensive pieces of quality steel.

American King Gillette's family had suffered in the Chicago fire of 1871, the fire that set back the fortunes of reaper inventor Cyrus McCormick (p. 94). Working as a travelling salesman to keep his family going, Gillette simultaneously hit on two momentous ideas: an improved safety razor, with only the very edge of the blade exposed, making skin cuts less common; and a new business model. By selling the device itself cheaply, good profits could be made by selling some element that required regular replacement. William Painter, inventor of the crown seal bottle-cap, reputedly told Gillette that a successful invention was one that would be bought over and over again by a satisfied customer. Gillette succeeded by making the relatively cheap steel blade disposable, designed to go blunt and be replaced after only a few uses. Many a manufacturer has since profited from the same philosophy.

Experts assured Gillette that steel blades that were thin, strong and cheap enough for his purposes could not be made. Gillette thought otherwise and, after six years of work and with the aid of young engineer graduate William Nickerson, he triumphed. His American Safety Razor Company, founded in 1901, went on to sell disposable razor blades by the billion. World War I provided a boost; the US government supplied his razors to every soldier. Sales soared further when changing fashions required women to shave their armpits.

In time, the company, now bearing this name, led the way in making the whole razor disposable, not just the blade, and in multi-blade razors. Gillette himself was financially ruined by the stock-market crash of 1929, an ironic fate since he was an opponent of rampant competition and capitalism, writing books promoting a socialist utopia.

Light from a Tube

Georges Claude, Peter Hewitt

The electric light globe, created by Thomas Edison and Joseph Swan around 1880 (p. 137), did not satisfy everyone as a source of light. 'Too small, too hot, too red' sniffed a rival inventor, hoping to do better. An alternative was already to hand. For decades physicists in Germany and elsewhere had been experimenting with 'discharge tubes'—sealed glass tubes with a metal plate (electrode) at each end. With most of the air pumped out and an electric current passing between the electrodes, the tubes glowed.

Replacing air with some other gas produced light of various colours. French physicist Georges Claude used the newly discovered rare gas 'neon'. This generated red–orange light, the archetypal 'neon sign'. He patented his tubes in 1902, publicly displayed them in Paris in 1910 and began selling into the booming US market in the 1920s. The tubes, now available in many colours, can be drawn out and shaped into letters and symbols as we see every day in advertising. The Claude Neon Company remains a manufacturer, though the industry grew up mostly by infringing Claude's patents.

American Peter Hewitt began using mercury vapour in discharge tubes in the 1890s. His lamps were less wasteful of electricity than incandescent globes, and produced lots of light, though it was an unattractive blue–green, which distorted other colours. That was less of a problem out-of-doors and mercury vapour lamps first reached the market in 1902, backed by George Westinghouse (p. 151).

By now the phenomenon of 'fluorescence' had been discovered. Certain chemical compounds (known as 'phosphors') glow when bathed in ultraviolet light. Such chemicals can make your washing look especially white and bright when hanging on the line in the ultraviolet light from the Sun. In time the technologies merged. A mercury vapour tube, coated on the inside with a phosphor, became the first fluorescent lamp (aka strip light). Their greater efficiency means 'light without heat' (or very nearly).

Pioneered by General Electric and first displayed at the New York World Fair in 1937, fluorescents are now available in various attractive colours. Since the 1980s they have been wound into convenient coils to replace ordinary electric globes, using only 20 per cent as much power and lasting much longer. Though 'cool tubes' now produce most of the world's artificial light, their days too are numbered. They will be increasingly replaced by light-emitting diodes or LEDs (p. 256) in the years ahead.

Hearts, Brains and Electricity

William Einthoven, Hans Berger

Italian anatomist Luigi Galvani discovered 'animal electricity' in the 1790s and so is the godfather of two key medical devices in the early twentieth century: the electrocardiograph (ECG) and the electroencephalograph (EEG). Both use electrical currents associated with nerve impulses to analyse the workings of our bodies, and built on a tradition of probing the human interior that started with the stethoscope (p. 87) and developed through to the use of X-rays (p. 156).

From the mid-nineteenth century, we knew that the heart of, say, a frog, contained electrical currents varying in synchrony with the pulse, and these could be detected with a sensitive meter. In 1903 Javanese-born Dutch engineer William Einthoven invented the even more sensitive 'string galvanometer' and used it to detect (and record on a paper chart) electrical currents picked up at various points on the patient's chest. Later models used the new electronic amplifiers (p. 176) to further increase sensitivity.

Einthoven identified half a dozen distinctive patterns of changing currents, some associated with a heart beating normally, others indicating abnormality and disease. His first machines weighed half a tonne and needed five operators, but a modern electrocardiograph (the name combines 'electricity', 'heart' and 'writing') fits easily on a tabletop. It may be combined with an oscilloscope (p. 161) to display the heart's electrical impulses on a screen, such as we find in every emergency ward. For his invention, Einthoven won the Nobel Prize for Medicine in 1924.

The human brain, centrepiece of the nervous system, also pulsates with detectable electrical currents. Here the pioneer was German physician Hans Berger. He had first tried to examine the workings of the brain by studying its blood flows, but in the 1920s turned to recording minute electrical currents detected with electrodes attached to the skull. His first electroencephalogram ('electricity', 'head' and 'writing') was of his own son in 1924. Studies by Berger and others soon revealed the variety of 'brainwaves', different patterns of varying currents, such as the 10 times-a-second alpha waves associated with relaxation.

Berger's invention of the EEG would later revolutionise neuro-medicine, as the ECG had revolutionsed the study of heart disease. Strict, reserved and increasingly isolated, Berger conducted his research secretly in his spare time, and gained recognition only in 1937. Despair at the rise of Nazism drove him to hang himself in 1941.

We Take to the Air

Wilbur and Orville Wright and Others

On 17 December 1903 American Orville Wright became first to fly in a heavier-than-air powered machine, while his brother Wilbur watched from the ground. At Kitty Hawk, North Carolina, the 'Flier', equipped with a 12-horsepower petrol engine, covered about 40 metres in 12 seconds. The Wright boys did not start from scratch. Any number of experimenters had sought the conquest of the air; the initial flapping wings, which never worked, were replaced by fixed ones able to glide as many birds do.

Around 1800 Englishman George Cayley began to study the forces that might help a human fly, summing up his ideas in the influential *On Aerial Navigation*. From 50 years of work, he knew the shape of the wings was critical in providing enough 'lift'. He added a tail for stability and tried a biplane design to add strength. He also knew lengthy flights would need some form of power. While his experimental craft never carried a human, many lifted the weight of one, showing it could be done.

After Cayley came many with the same desire. The German engineer Otto Lilienthal boldly went aloft in his own machines; his 1889 book on aerodynamics set the Wright Brothers on their course. In 1896, after 2500 flights, Lilienthal lost control of his glider in a strong wind gust, crashing to his death. American astronomer Samuel Langley, Secretary of the Smithsonian Institution in Washington, tried adding a steam engine to a glider. His first model flew a kilometre before running out of fuel; a later, much larger, version proved too heavy to fly. Langley gave up the hunt.

Many experimenters were now at work, achieving long hops across the ground, if not sustained powered flight. The American Octave Canute summed up the situation in his 1894 *Progress in Flying Machines,* drawing on all the pioneers including the Australian Lawrence Hargrave, for whom who he had a particular regard. The book, and Canute himself, was another source of inspiration for the Wrights.

The big day at Kitty Hawk climaxed half a decade of methodical effort by Wilbur and Orville, printers and bicycle builders from Dayton, Ohio. They had worked their way up from kites through gliders and on to various models of powered aeroplane. In a key development, their craft could be controlled by the pilot about all three axes, turn, pitch and roll. Taking turns, they made four flights that first day, the longest lasting nearly a minute and covering 250 metres, before the plane was overturned and wrecked by a sudden wind gust. For two years they continued to work in secret, steadily increasing the distance and altitude of their flights, holding back public demonstrations until 1908.

By then, daring inventors in several countries were taking to the skies. Aided by new designs and lighter petrol engines, their creations could stay aloft for half an hour and cover many kilometres, though the engines and the craft themselves remained unreliable and the flights literally death defying. An early trophy was claimed in 1909 by French aviator (and successful maker of automobile headlamps) Louis Blériot—a £1000 prize offered by the London *Daily Mail* for the first crossing of the English Channel. Reaching Dover in 37 minutes, he delighted his countrymen but worried the British. The bulwark of the English Channel, which had halted the invasion plans of the Spanish Armada and of Napoleon, now seemed less secure. Politician David Lloyd George said with some concern 'Flying machines are no longer toys'.

World War I, erupting only five years later, showed the truth of that. Aircraft proved powerful in combat, for reconnaissance and for bombing, roles originally assigned (by the Germans at least) to airships (p. 165). The combatant nations established embryonic air forces. Aircraft fought each other in aerial 'dog fights' over France and elsewhere; a later innovation allowed them to shoot their machine guns through the whirling blades of the propeller, increasing accuracy of fire. To survive such scraps, planes needed to be stronger, faster and more manoeuvrable—the science and technology of aircraft advanced rapidly. Without the stimulus of military need, aviation may have been much slower to take off.

In 1919 that military influence was played out in peacetime. British aviators John Alcock and Arthur Brown flew a Vickers Vimy bomber, stripped of its wartime accessories and fitted with extra fuel tanks, in the first-ever crossing of the Atlantic by airplane. In quest of another *Daily Mail* prize, this time £10 000, they took 16 hours to fly from Newfoundland to Ireland, landing in a bog.

With the long-distance capacities of aircraft demonstrated, the way was open for regular flights to carry mail, cargo and, soon, paying passengers, all within 20 years of the Wright Brothers' first achievement. Much more lay ahead.

Rockets, War and Space

Konstantin Tsiolkovsky, Hermann Oberth, Robert Goddard

The intercontinental ballistic missile (ICBM) unites two technologies that grew up during World War II, the atom bomb (p. 216), now in the form of the fusion-powered 'hydrogen bomb', and liquid-fuelled rockets, greatly progressed by German engineers. The combination is perhaps the most terrifying weapon ever devised, able to send the devastating force of a nuclear warhead many thousands of kilometres in a matter of minutes.

The invention of gunpowder 2500 years ago by the Chinese led to its use in 'fire arrows'—tubes of gunpowder strapped to the shafts of arrows and ignited. More powerful devices, with shrapnel-filled explosive warheads, were used in the thirteenth century to battle Mongol invaders. Gunpowder has its limitations. Just as its place in mining and cannons was taken in the nineteenth century by other explosives, liquid fuels supplanted it in rockets, making them both more powerful and more controllable.

Three great names illuminate this story: Russian Konstantin Tsiolkovsky, German Hermann Oberth and Robert Goddard of the USA.

Tsiolkovsky, scientist and schoolmaster, was a visionary and the 'father of astronautics'. In 1903 he proposed using rockets to begin the exploration of regions beyond the Earth. If a rocket could reach a high enough speed, about 11 kilometres per second, gravity could not stop it escaping into space. The rocket's ultimate speed depended on how fast gas particles were ejected from its rear end; liquid fuels like kerosene or alcohol would produce faster moving exhaust gases that any known solid fuel. He also proposed the multistage rocket—several rockets, one atop the other, each firing in turn as the one below exhausted its fuel and was discarded.

Transylvanian-born Oberth joined the enterprise in 1923. Aged only 19, he wrote the brief but influential *Rockets into Interplanetary Space*. This and later writings spurred the development of the Society for Space Travel. By the early 1930s society members were launching liquid-fuelled rockets. Although often erratic, these could reach a few kilometres in altitude, well short of space flight.

Robert Goddard had already launched a rocket powered by gasoline and liquid oxygen in 1926, though it rose a mere 12 metres from the ground. A decade later, his creations could reach an altitude of 2 kilometres and top speeds of 700 kilometres per hour. His optimistic book on the subject, *A Method of Reaching Extreme Altitudes*, was published in 1919.

Further progress, especially in Germany, was spurred by the possibility of military use. With research funds always short, space-travel enthusiasts won support from the army. By 1937 a research station near the village of Pennemunde on the Baltic coast, under the youthful Werner von Braun, was testing fuels like alcohol, hydrazine, liquid oxygen and nitric acid. This culminated in the A4 rocket, which could reach an altitude of 100 kilometres, a range of 1000 kilometres and a top speed of 5000 kilometres an hour. As the V2, one of Hitler's 'terror weapons', this bombarded London and other centres late in World War II—too late to make a difference.

The post-war world was more profoundly affected. Von Braun and other German experts were recruited by the victors of the war, most going to the USA, some to the USSR (now the Soviet Union). With the USA and the USSR now on opposite sides of the new 'Iron Curtain', rising tensions drove further developments. Rockets became more powerful and reliable, equipped with guidance systems to direct them more accurately to targets even on the other side of the world, thanks also to the smaller weight of more advanced warheads.

The arms race now extended into space. Both sides sought to demonstrate the power of their missiles (and economic philosophies) by placing packages of instruments, and later humans, into orbit around the Earth. The Soviets achieved the early successes: the first satellite (*Sputnik One* in 1957), the first man into space in 1961, the first spacecraft to leave Earth's orbit, the first images of the far side of the Moon, the first spacewalk and many others. The US effort had been fragmented within the military. Not until the space agency NASA was founded in 1958 was it sufficiently unified and focused to match the Soviets.

By the 1960s the goal was 'a man on the Moon', achieved by NASA in 1969. However, earth-orbiting satellites had found many other uses, for instance in communications (p. 257) and weather forecasting. The hunt had begun for reusable systems to replace the use-once-and-discard rockets that had powered the Space Age until then.

Radio Valves that Amplify and Oscillate

Lee de Forest

Electronics, with all its immense potential, had begun with Englishman John Ambrose Fleming's 'radio valve' (p. 140), patented in 1902. This addressed one of the big challenges with the still embryonic technology of wireless, how to best to 'rectify' the oscillating signals from the receiving antenna so the message they carried could be made audible in headphones.

Other challenges remained. The father of American electrical engineer Lee de Forest had hoped his son would follow him into the Congregationist ministry but young de Forest had been interested in electricity and wireless from his youth. In a crucial development, de Forest added a third electrode to Fleming's two-electrode diode in the form of a wire mesh or 'grid' between the heated filament (cathode), which gave off electric current as electrons, and the cold plate (anode) which received them.

By varying the current in this 'grid', he could control a much bigger current flowing through from cathode to anode. He could amplify weak signals into much stronger ones.

This immediately interested Western Electric, part of the Bell Telephone (AT&T) empire. Bell desperately needed the new valve, dubbed the 'audion' and patented in 1906, to amplify telephone signals, which inevitably weakened as they travelled down the wires; This would make long-distance telephone calls possible, and the telephone business more profitable. It was Western Electric that turned de Forest's baulky prototypes into cheap, reliable and long-lived devices with a multitude of uses.

De Forest's creation could do still more. Connecting the grid and plate together with some external circuitry made the valve oscillate, generating the high-frequency alternating currents needed to transmit wireless signals but which until now had been made by electrical sparks (p. 160); wireless operators on ships were all known as 'Sparks'.

Able both to amplify and to oscillate, the triode (the name came from its three electrodes) was a vital advance. For more than 50 years, radio, television, radar, sound recording and even the first computers would all depend on de Forest's invention and its descendants.

De Forest's achievement was not his alone. Austrian Robert von Lieben had come up with the idea of a three-electrode valve at the same time. De Forest lived to be nearly 90, but von Lieben died in 1914 aged only 36, too young to take his technology on and to play his proper role in the new age of radio.

Music in the Air

Reginald Fessenden

In the century-long history of radio, the name of Canadian-born American Reginald Fessenden is not well known. Yet Fessenden was a real pioneer, broadcasting music and speech by wireless when everyone else, including the great Gulielmo Marconi, was still tapping out morse code. The head of the General Electric Laboratories later called him 'the greatest wireless inventor of his age, even greater than Marconi', which is quite a call.

Fessenden's greatest triumph was to outdo Marconi, who in 1901 had sent the first morse code signal across the Atlantic. Unlike Marconi and most others, Fessenden believed that wireless waves could also be imprinted with speech and music. Very few people agreed with him.

On Christmas Eve 1906 he proved it so. Radio operators on ships off the coast near his laboratory on Brant Rock, Massachusetts, were amazed when their headphones delivered not the familiar dots and ashes of morse code, but Handel's *Largo,* reproduced from an Edison cylinder, and the man himself playing *Silent Night* on the violin and reading from the Bible, before wishing his listeners 'Merry Christmas'.

Fessenden had used a high-speed electrical generator (p. 95) to produce the high-frequency oscillating currents he needed to generate radio waves in his transmitting antenna, rather than the commonly used, but crude, spark gaps. Nonetheless, most people, even Essendon's own backers, believed Marconi: such a broadcast, even if feasible, was of no value or interest. More than a decade went by before 'radio broadcasting', rather than mere wireless telegraphy, began its phenomenal rise in popularity (p. 189).

Fessenden was 15 years ahead of his time and prolific. His 500 patents included a depth sounder for shipping, using sound waves (this gave him a comfortable income); signalling systems for submarines; and a smoke generator to hide tanks on the battlefield. Early in his career, he had worked as a chemist for Thomas Edison and later for George Westinghouse, helping him to light the 1893 Columbian Exhibition with electricity. Fessenden proposed very early what was later called 'radar'. He argued that a system that detected objects like icebergs by reflecting radio waves off them could prevent disasters. Again he was ahead of his time: 20 years passed before radar was ready for use (1935). Fessenden did not live to see it; he died from heart disease in 1932, aged 62.

A Better Compass

Elmer Sperry

The magnetic compass—invented by the Chinese, enhanced in Europe and used by every mariner in the great age of exploration—has limitations. It does not point naturally towards the north and south geographic poles but towards the slightly different (and slowly moving) magnetic poles. You cannot trust it absolutely.

Its modern replacement is the 'gyrocompass', invented by several people but most conclusively by the American Elmer Sperry in 1908. He started with the gyroscope, invented in the previous century by Frenchman Leon Foucault. Any spinning object tends to stay spinning with its axis pointed in the same direction, a manifestation of 'rotational inertia'. So a spinning bullet from a rifle does not tumble, and a moving bicycle stays upright on its two narrow wheels. A gyroscope is a metal ball or disk turning at high speed (nowadays driven by an electric motor) and so able to stay pointing at the same spot in the sky. Three gyroscopes at right angles keep the Hubble Space Telescope directed very precisely at whatever distant object it is photographing.

If you push and pull on the opposite ends of the gyroscope axis (so applying a 'torque' or 'couple') in an effort to make it change direction, it responds in an odd way. It turns itself at right angles to both the direction of its own axis and the direction of the applied torque. If the torque is the result of Earth's gravity, the gyroscope will slowly shift until its axis is pointing genuinely north–south, along the Earth's axis, and there it will stay. The gyroscope becomes the gyrocompass. Nowadays all well-equipped airplanes and ships have one, and usually more than one, since the movements of the 'master' gyrocompass can be transmitted to repeaters in other locations. Gyrocompasses copes are also totally immune to the effects of iron and steel nearby, another problem with the magnetic compass.

Sperry was typically prolific. He held 400 patents when he died in 1930, including for machines and fuse wire. His gyrocompass was first installed in the US battleship *Delaware* in 1911. It had many uses in peace and war, including gyroscopic gunsights and bombsights; these kept pointing at the target however the plane or ship moved. The Sperry Corporation, founded in 1933, is now part of Unisys.

Pioneers of Television

Shelford Bidwell, Paul Nipkow, Alan Campbell-Swindon

If we go looking for the pioneers of the now ubiquitous technology of television, we can start with the Italian-born priest Abbé Caselli. By 1862 he had a system to send pictures by wire. The results were crude, the images little more than shadows. Nonetheless, Caselli quickly gained the useful patronage of Napoleon III and set up a number of stations across France for sending hand-written messages and drawings over telegraph wires.

The venture never made money. Interference was one problem. Dots and dashes from Morse Code messages sharing the same line marred the images. Of course the good Abbé's pictures did not move. His technology was more like a facsimile than television as we know it now. And few would have expected otherwise. The first 'moving pictures' were still 30 years away (p. 158).

The Two Problems

There were always two challenges with 'television'; how to break a picture up into elements which could be transmitted, and how to convert those fragments of image into an electric current which would pass down a wire. In 1873 there was an answer to the second. French physicist Henri Becquerel (later to discover radioactivity) found that the newly discovered element (technically a semiconductor) selenium could be used to make a 'photo-electric cell' to turn light into electricity. The brighter the light falling on the cell, the stronger the current it produced.

As for breaking the picture into fragments, the English inventor Shelford Bidwell tried moving a selenium cell across the face of a picture using cams. That was in 1881. Three years later, the German Paul Nipkow had a much better idea, a concept used in some of the first television transmissions 50 years on. Nipkow punched a series of holes arranged in a spiral pattern into a metal disc. He placed his 'Nipkow disc' between the picture and the photo electric cell and spun it rapidly. This broke up the picture, as seen by the selenium cell, into many dots; the brightness of each dot controlled the amount electric current leaving the cell.

The Bidwell Vision

In 1908, Bidwell proposed in a letter to the science journal *Nature* what he called 'electric vision'. It called for 90 000 selenium cells in a grid, each looking at one segment of an image, each connected by its own wire to a similar point in the receiver. The breathtaking impractability of this scheme made Scottish-born electrical engineer Allan Campbell-Swinton wonder the same result could somehow be achieved with just a single wire between transmitter and receiver.

Campbell-Swinton was highly talented. He had taken some of the earliest X-ray photographs, and helped the struggling Marconi (p. 160) by writing a letter of introduction to the head of the British Post Office, Sir William Preece. Most importantly, he knew that the German scientist Karl Braun had invented the cathode ray tube or CRT (p. 161), in which electric and magnetic fields could move an electron beam across a surface inside an evacuated glass container.

The Campbell-Swinton Response

Campbell-Swinton reacted to Bidwell's 90 000 cell proposal in his own letter to *Nature*. He argued the Braun tube could make electric vision possible by serving as both transmitter and receiver, as both camera and screen. At one end of the system, light from the image or object would fall on a selenium plate segmented into many pieces, with each cell of the plate accumulating electric charge in proportion to the intensity of the light falling on it. Scanning an electron beam across the face of the plate would pick up the accumulated charge cell by cell, resulting in an electric current which varied in synchrony with the changing brightness of the scanned image.

The current, passed down a wire to a similar tube some distance away, would cause changes in the strength of a second electron beam, causing it to paint in the picture on the face of the CRT tube as a pattern of dots of fluctuating brightness. The slowness of the human eye would do the rest, as indeed it did with the illusion of movement created by the relatively new 'moving pictures'. If the whole process could be repeated quickly enough, the images provided by 'electric vision' would also move.

In 1911, Campbell-Swinton patented this idea. 'It is just an idea', he wrote in his application, 'and the apparatus has never been built. Furthermore, I do not for one moment suppose that it could be made to work without a great deal of experimentation and probably much modification'.

It did take a quarter of a century of work to produce a really workable system based on his idea (p. 206), but Campbell-Swinton's insight was nonetheless brilliant. He had anticipated the form and function of the cameras and display systems that formed the heart of video technology almost to the present day.

Ammonia in War and Peace

Fritz Haber

When the German chemist Fritz Haber won the Nobel Prize in 1918, many people were alarmed. Haber had promoted a new and terrible weapon for the trenches of the Western Front—poison gas, initially chlorine, first used at Ypres in 1915. His wife, herself a chemist, had already reportedly committed suicide over the matter.

Haber's Nobel Prize had nothing to do with poison gas. Instead, it acknowledged his process to make the vital chemical 'ammonia' synthetically, by combining nitrogen and hydrogen under the right conditions of temperature and pressure, with a catalyst to greatly speed the process. The ammonia, oxidised to nitric acid, one of the most important industrial chemicals, could then be used in explosives and fertilisers.

Germany, like most industrial nations, depended on imports of the mineral salpetre from Chile and on by-products from making coal gas. Fears were growing that the rising demand would see supplies of the first would soon run out, with the latter unable to make up the difference. The atmosphere of course could provide an almost unlimited supply of nitrogen, but some 'fixing' process was needed to form it into useful compounds. Haber invented that process, and another leading chemist, Carl Bosch, from the company BASF, refined and upgraded it for major industrial use.

The invention of the Haber–Bosch process may have helped precipitate World War I by encouraging the German government to think it could be self-sufficient in ammonia and its products, and so sustain its industry through war and the expected trade blockades. At least it probably prolonged the war by maintaining German supplies which would otherwise have been exhausted. More than a century earlier, the French government had called for new methods to make soda, another important but vulnerable chemical, resulting in the Leblanc process and its aftermath (p. 67).

With modifications, the Haber–Bosch process still produces ammonia for synthetic fertilisers today. So its impact in peace is perhaps even greater than in war, empowering, for example, the 'Green Revolution'. In wider industry, the use of high pressures and catalysts to speed up chemical processes, as Haber did, has helped the refining of petroleum and the production of oil from coal.

Haber had a broad range of interests and achievements. One failure was an attempt to recover gold from seawater to help Germany repay its war reparations. He died in Switzerland in 1934, exiled from Germany because of his Jewish background.

Automobiles by the Million

Henry Ford

American Henry Ford transformed the motorcar from a luxury item (p. 146) into a workaday machine almost anyone could own, and implanted the 'assembly line' firmly in the lexicon of manufacturing. The Ford Foundation, created from his immense fortune, currently disperses half a billion philanthropic dollars a year. It is a heady legacy for the dour Michigan farm boy, a tinkerer with machines since his youth.

Ford built his first car in 1896, using four bicycle wheels. In 1903, aged 40, he quit working for Thomas Edison to found his car-making firm, declaring he would make 'a car for the great multitude', a promise he kept. The first of the iconic Model T Fords came off the production line in 1908, costing $1000. Over two decades he built 15 million of these pioneering machines; the price fell 75 per cent.

The result was an irrevocable change in American (and later global) society. Cheap personal transport transformed the shape and operation of cities, creating the suburbs (and later gridlock), linking widespread population centres, enriching lives at a cost to the environment (though the first cars promised relief from streets knee-deep in horse manure).

Ford and his son Edsel made many improvements to automobiles (including the one-piece V8 engine in 1932 to drive a new generation of large, powerful and smooth-riding cars), but his greatest innovations were in the methods of manufacture. He did not invent mass production or the assembly line; they were already in use making guns, typewriters and sewing machines; but brought them to a state of exquisite efficiency. His Highland Park plant in Michigan divided production into dozens of simple steps, undertaken at a relentless place. As partly completed automobiles moved down a 300-metre conveyor belt, components were added by workers, each doing only one task repetitively, in a time dictated by the movement of the belt. At peak production, around 1914, a 'Tin Lizzie' took only 90 minutes to traverse the line, a completed vehicle emerging every 30 seconds.

These immense gains in productivity allowed Ford to treat his workers relatively well, playing double the wages on offer elsewhere for fewer hours. But they were essentially slaves to the line, which could be speeded up to further enhance production and ran 24 hours a day with three shifts. Though Aldous Huxley later made 'Our Ford' a subject of veneration in *Brave New World* (p. 121), the endless repetition of simple tasks was satirised by Charlie Chaplin in *Modern Times*.

Creating
Chemical Cures

Paul Ehrlich

When William Perkin found a beautiful purple dye in coal tar, previously merely a waste product from making coal gas and coke (p. 114), he started a gold rush among chemists. They began systematically to identify all sorts of chemical compounds, many of them organic (containing carbon). As methods improved, thousands were extracted and named. Since these chemicals, or at least the possibility of them, had always existed, we cannot say most were invented. Innovation came through finding uses for them in industry or medicine—as dyes, drugs or synthetic materials to be manufactured and sold, and through careful reworking of the compounds to maximise their effectiveness.

The vision that some of these new chemicals might help to control disease and promote health motivated Paul Ehrlich, a German doctor who spent his early career working with Robert Koch, discoverer of the microbes that cause anthrax and tuberculosis (p. 143). By the early twentieth century, Ehrlich was deeply involved in the search for chemicals that could be targeted against specific diseases or conditions without harmful side effects—'magic bullets' as he called them. Hundreds of chemicals were tested in search of those that worked. An early target was a drug to treat sleeping sickness, caused by a parasite in the blood that could be killed with the trypan dye.

Ehrlich's public fame rests mostly on salvasan, the arsenic-containing chemical that killed the organism that caused syphilis. The successful drug was number 606 on his list of trials (though it had been initially judged ineffective) and number 914 (neosalvarsan) proved even better in use. Ehrlich, who ate little, was sustained through these labours by a diet of 25 strong cigars every day; he always had a cigar box under his arm.

Ehrlich had to battle opposition to his new ideas and methods, but he persevered, benefiting public health and cementing his own reputation; he already had shared the 1908 Nobel Prize for Medicine. The 'chemotherapy' he pioneered is a now vast enterprise and industry. The invention and marketing of chemicals as pharmaceuticals for almost every ailment, physical and psychological, known to the human race, has been made possible due to better understanding of how the structure of the chemicals allows them to interact with the chemical workings of the body (p. 222).

Making the Traffic Flow

Lester Wire

Traffic lights, now essential to keep traffic moving smoothly in our city streets, are a technology with many claims to paternity, making it hard to trace the story with certainty. If we mean the present red-yellow-green system, its first appearance was in New York in 1918, or perhaps Detroit in 1920. In all probability, it was in the USA and in response to the growing volumes of traffic as the factories of Henry Ford and others poured out motorcars cheap enough for the average person to buy (p. 182).

Attempts to control traffic by signals dated back at least 50 years earlier to a signal post with movable arms like railway semaphores set up to outside parliament house in London in 1868. Red and green gas lamps provided warnings at night, although according to reports the signal blew up after a year of use, killing a policeman and discouraging further experiment.

The first electric traffic signals, though with only red and green lights, were the work of Salt Lake City policeman Lester Wire in 1912. Placed in charge of the traffic squad, he sought to ease the chaos on the city roads caused by the increasing popularity of automobiles. He also wanted to save fellow policemen from standing in the rain to direct traffic. Instead they could control the lights from a shelter nearby. Wire refined his system but never patented it, and died having received no royalties.

In 1923 Afro-American inventor and sewing-machine repairman Garrett Morgan, also the inventor of a gas mask and a chemical to straighten hair, patented a cheap and simple traffic control system, using hand-cranked movable arms. This became quite popular, and the myth grew that Morgan had invented traffic signals, which is clearly not so. The US patent office had issued 50 patents before his.

Early systems could be operated either automatically, the lights changing at set time intervals, or manually, controlled by a policeman or other official able to view the state of the traffic and make appropriate adjustments. Not until the 1960s were electronic circuits inserted in the roads to monitor traffic flows, so that the cycling of the lights could be adjusted to minimise delays. From the 1980s, information technology allowed a variety of light-change cycles to be applied at any particular intersection as traffic conditions changed, and the succession of lights down a stretch of road to be coordinated.

Steel that Does Not Stain

Harry Brearley

The invention of stainless steel in 1912 by British chemist Harry Brearley shows how an invention can end up being used for very different purposes than those originally intended. At the time, Brearley was in charge of a laboratory set up by two of the major steelmakers in Sheffield in the UK, Firths (by whom he was employed) and Browns, to solve various problems the industry faced. Such collaborative research was a new idea.

A small-arms manufacturer had reported that the steel barrels wore away quickly from the effects of the hot gases released. After many trials, Brearley found that adding about 10 per cent of the metal chromium to the steel greatly reduced wear. Problem solved. Needing to etch samples of the steel for study under the microscope, he found the chromium-enriched steel also resisted attack by acids, including common food acids like vinegar and lemon juice. Brearley was quick to see the potential for cutlery manufacture, a major business in Sheffield. Much cutlery was electroplated with chromium or nickel already and so resisted corrosion, but plated steel would not take a good edge. In consequence, knife blades had to be made from mild carbon steel, which quickly rusted or stained unless less well cared for. Knife blades of the new 'rustless' steel would be largely immune to staining and other corrosion.

Sheffield cutlery makers were conservative. Brearley had to have samples made with the new steel by the local cutler, Mosley's, at his own expense; the manager at Mosley's suggested calling the new material 'stainless steel'. Slowly its advantages were accepted. The major German steel manufacturer Krupps obtained a similar result using nickel; the steel produced was softer and more easily worked and even more resistant to acids. The outbreak of war halted production, but in the 1920s the use of chromium and nickel together was pioneered and remains the basis of stainless steel today.

Brearley soon fell out with Firths. They had made his services available to run the laboratory, but claimed ownership of the invention because he worked for them. Brearley joined rival company Browns to continue the development of stainless steel. Having left school at 12 to wash bottles, Brearley was largely self-taught in steel matters, but his expertise was recognised in 1920 by the Iron and Steel Institute, which awarded him the Gold Medal named after the great steel pioneer Henry Bessemer (p. 116).

The Games People Play

Arthur Wynne, Clarence Darrow

Crossword puzzles, popular everywhere, are a recent invention. Some very simple ones appeared in the nineteenth century, but the puzzles' profile today stems from 1913, when English-born journalist Arthur Wynne published one in the *New York World*, a Sunday newspaper. As today, clues were given to identify words that fitted together vertically and horizontally, but Wynne's first puzzles had no black squares. The outbreak of war diverted attention to other matters, but the 1920s saw an explosion of interest, with crossword puzzles (initially called 'word-cross puzzles') appearing in British newspapers from 1922 and in *The Times* from 1930.

The arrival of electronic games might have seen the end of the crossword, but in fact information technology now supports the pastime, with computer programs available from the late 1990s to allow anyone to construct a puzzle for themselves. An estimated 40 million people in the USA alone wrestle with a crossword daily.

If the crossword is the most popular pen-and-paper game (though a challenge has recently come from Sudoku), Monopoly remains the best-selling board game, despite the rise of Trivial Pursuit. With 200 million sets sold in 20 languages, Monopoly has made a fortune for its publishers, Parker Brothers, but it nearly was not so. Charles Darrow, an unemployed salesman from Germanton, Pennsylvania, devised the game in the early 1930s for his own amusement, based on Atlantic City and drawn on his tablecloth. The game became popular among Darrow's family and friends and he hand-made copies to sell. Surprisingly, an approach to Parker Brothers was rejected. They claimed to find 52 'design errors', such as that the game took too long and the rules were too complex.

Undeterred, Darrow continued to market Monopoly himself, having a printer produce 5000 copies. Parker Brothers soon saw their mistake and offered Darrow a royalty deal. In its first year, 1935, Monopoly was immediately the most popular game sold in the USA, and Darrow was soon the first board-game millionaire.

Darrow was not in fact the first to design such a game, just the first to make a real success of it. Half a dozen board games in which property or shares were bought and sold, such as Lizzie Magie's The Landlord's Game, were in existence, some patented. Some may have influenced the game Darrow devised, even its title. Parker Brothers quickly bought up the rivals, some very cheaply, to protect its investment.

Women Escape the Corset

Marie Tucek, Mary (Polly) Jacobs

In their attempts to conform to the fashionable body shape, for centuries women forced themselves into uncomfortable, even dangerous, undergarments. The corset (from old French for 'small body') was reputedly invented by the French Queen Catherine de Medici in the 1550s, to ensure her ladies had narrow waists. Rods of whalebone or even metal were sown into the garments, and the restraint enforced by lacing up the back.

By the late nineteenth century, women began to rebel against the ordeal of the corset, and inventive minds were devising alternatives. Any number of women, most of them French, helped invent the brassiere; some took out patents. Marie Tucek designed one in 1893, looking much like a modern bra, with separate pockets for the breasts, and over-the-shoulder straps ('brassiere' is Old French for 'upper arm'; the breasts being supported from above rather than squeezed up from below, as in the corset). But Marie failed to market her invention successfully.

Across the Atlantic in 1910, New England socialite Mary (Polly) Jacobs sewed together two silk handkerchiefs and some ribbons to make a suitable undergarment for wearing beneath a sheer and revealing gown. She took out a patent in 1914 and started selling her invention under the business name of Caresse Crosby. Though soft and light to wear, the Jacobs bra was more flattening than flattering (though that proved to be the preferred style in the Jazz Age of the 1920s). No businesswoman, Jacobs soon sold out to a corset company called Warners, which was to make 1000 times the patent purchase price over coming decades.

World War I speeded the rise of the bra. With their menfolk at sea or at the Front, women were taking their places in factories and offices and needed comfortable undergarments. In 1917 the US War Industries Board entreated women not to wear steel-reinforced corsets for the sake of the war effort. Women responded, reputedly liberating steel sufficient for two battleships.

Through the 1930s and 1940s, new designs for brassieres emphasised uplift, as the corset had. Some major engineering minds were at work, including the aircraft builder Howard Hughes; his ingenuity emphasised the appeal of actresses like Jane Russell. In 1928 Russian immigrant Ida Rosenthal of the company Maidenform defined cup sizes, so bras fitted better. Specialised bras for nursing mothers came in the 1930s; in the 1970s the 'sports bra' successfully combined restraint with comfort.

The Tank Goes to War

Ernest Swinton, Leslie Skinner

The military tank was invented more than once. Leonardo da Vinci drew one around 1500, and the Australian inventor Edward de la Mole designed one driven by a petrol engine in 1912. Neither was ever built, largely because no one saw the need.

Trench warfare in World War I saw calamitous losses but no sustained progress by either side. Tens of thousands were killed in futile advances over chaotic broken ground, through barbed wire (p. 132) and in the face of withering enemy fire, particularly from machine guns (p. 101). British colonel Ernest Swinton glimpsed a solution in an article on American farm tractors that ran on caterpillar treads. Those could handle the battlefield terrain, he thought, and covered with armour-plate and equipped with guns, would make an unstoppable advance through 'no man's land', a phrase he had invented as an official war correspondent.

He built a primitive prototype that impressed military authorities, including Winston Churchill, and the first vehicles were ordered. Production began in 1916. Mark 1, called Big Willie, was box-shaped, weighed 26 tonnes and had two side-mounted cannons, six machine guns and a crew of eight. With secrecy paramount, workers were told they were building 'water carriers for Mesopotamia', later simplified to 'tanks', the name that persists today.

Tanks first went into battle in September 1916, in an effort to break the murderous stalemate of the Battle of the Somme. Forty-seven tanks began the advance at dawn, terrifying the Germans as they emerged, rumbling relentlessly, through the smoke and fog. Mechanical failures and fuel shortages soon halved the force, but the remaining tanks penetrated 10 kilometres through German defences before withdrawing. That first encounter was far from decisive, and the war would still run another two years, but the potential of the weapon was clear. Other versions soon followed. The Whippet could reach 16 kilometres per hour, the Hornet had a turret for all-round fire.

Much enhanced, tanks have remained prominent in land warfare ever since, though not without challenge. In 1940, seeking a decisive anti-tank weapon, US Army Colonel Leslie Skinner put an explosive charge on top of a small high-speed rocket and fired the combination from a metal tube that could be held over the shoulder. The Mark 1 Rocket Launcher was soon renamed the Bazooka after a bizarre homemade trombone played by popular comedian Bob Banks.

The Birth of Radio Broadcasting

David Sarnoff

Many of the names associated with great inventions are not the inventors, but the visionaries who can see where the invention might lead. In the case of radio, David Sarnoff was that man. Following Guglielmo Marconi's first signal across the Atlantic in 1901 (p. 160), the new technology of 'wireless' slowly made its mark. Early use concentrated on messages between ships, or from ship to shore, replacing flags and flashing lights. When two ships collided in fog off New York in 1900, a message from the 'Marconi' operator on one ensured the passengers were quickly rescued.

Longer range wireless telegraph services soon followed. In 1903 the US president and the British king exchanged telegram greetings across the Atlantic. Reports of the attack on Port Arthur during the Russo-Japanese war in 1905 came in by wireless. Explorer Robert Perry sent a telegraph message from the North Pole in 1909. A year later, radio aided the capture of escaped British murderer Robert Crippen while he was still at sea (p. 97).

The eruption of war in Europe in 1914 provided an immense stimulus. Ground controllers could direct airplanes and airships (more new technology) to their targets using radio; quick warnings could be given of gas attacks. Intercepting radio signals enabled one side to know the other's plans in advance; the 1917 Zimmerman Telegram that brought the USA into World War I is the most notorious example. Increasing power and sophistication matched growing use; better valves derived from de Forest's triode (p. 176), new radio circuits improved reception.

When peace came in 1918, people thought about using radio as a source of entertainment and information. David Sarnoff, working for the Marconi Company, had outlined his vision in 1916—a 'radio music box' in every home, broadcasting music, educational lectures, news reports and sports scores across the country. People in remote areas would particularly benefit. Sarnoff was later to head up the Radio Corporation of America (RCA), a leading player in radio (and television) and founder of the NBC network.

By 1920 Sarnoff's vision was taking shape. Radio broadcasting as he had envisaged it began in November that year, when radio station KDKA went to air from the Pittsburgh garage of an employee of the Westinghouse Corporation, bringing regular broadcasts of music and news. One of its first reports told of the election of Warren Harding as US president. Many others followed. We had entered the age of radio.

Dating with Radioactivity

Arthur Holmes, Willard Libby

The discovery of radioactivity in 1896 transformed a long-running debate about the age of the Earth. Heat released by radioactive minerals in rocks meant that the Earth could be hundreds of millions of years old and still not have gone cold, enough time for biological evolution and massive changes in the landscape. And the possibility of 'radioactive dating' opened up. The slow decay of radioactive minerals, one into another, could serve as clocks to reveal just how old the Earth is.

Some radioactive elements had offspring that decayed quickly, while uranium, thorium and radium took billions of years. A measure of the passing ages might come from comparing the proportions of long-lived and short-lived radioactive elements in a mineral sample.

Such measurements were complex and time-consuming; many geologists thought them not worth the trouble. But Englishman Arthur Holmes persisted with the technique through to ultimate triumph, battling indifference, refining his techniques and reducing the errors. By 1921 he was winning. Speakers at the British Association meeting that year agreed that radioactive dating was valid and that it indicated the Earth was 'several billion' years old.

For much more recent dates, we use 'radiocarbon dating'. The brainchild of American chemist Willard (Frank) Libby, this is now used to date all sorts of ancient remains and artefacts up to 60 000 years old. Aged only 31 when war broke out in Europe, Libby joined a team separating the isotopes of uranium so an atomic bomb could be made (p. 216). After the war, radioactivity continued to play a major part in his career.

Radioactive carbon 14 (C14), discovered in 1940, constantly decays into something else, giving off detectable radiation. Like ordinary carbon, C14 (as carbon dioxide) is absorbed from the air into plants. When a plant (or a plant-eating animal) dies, C14 is no longer taken in. By measuring the amount of C14 remaining in something once living—say, a piece of wooden furniture, ashes from a wood fire, a piece of old parchment or cloth, an Egyptian mummy—Libby could tell how long it was since the plant involved died.

Libby won the 1960 Nobel Prize for Chemistry for this invention, which was of immense value to archaeologists and historians. Carbon 14 has a 'half-life' of 5700 years. After 10 half-lives (57 000 years), only a millionth of the original radiocarbon remains. That puts a limit on how far back we can go with radiocarbon dating; and radiocarbon dating is good only for once-living materials, not minerals.

Duplicating Signatures, Detecting Lies

John Larson, William Marston

The term 'polygraph' means 'multiple writings'. In 1804 US President Thomas Jefferson proclaimed one polygraph, the work of English-born Isaac Hawkins, the 'finest invention of the present age'. Well he might, since it let him sign more than one official document at a time. Two or more pens were connected by a system of jointed rods. The movement of one guided by the hand was accurately and instantaneously reproduced by the others. It is not a device in great demand nowadays.

The better known polygraph is the 'lie detector'. The multiple writings here continuously record things like breathing rate, blood pressure, pulse or perspiration on the skin. Advocates argue that these responses will automatically reflect a state of stress, brought on by telling an untruth. For example, by 1895 Italian criminologist Cesare Lombroso was recording changes in the blood flow of someone undergoing an interrogation. Over the next few decades other responses were thought a guide to the truth or otherwise of answers to question. One machine monitored skin sweatiness by measuring its electrical conductivity. The liar was thought to sweat more, and the saltiness of sweat makes it conduct more current. Another machine measured significant changes in breathing, using an air-filled tube wrapped around the chest.

Pulling these together into one machine was the work firstly of John Larson, a medical student working for the police department in Berkley, California in 1921, and later by Leonard Keller of Chicago and Dr William Marston of Harvard (the latter's diverse achievements included creating the comic book heroine Wonder Woman). Marston in particular realised the machine was only part of the lie-detection process. Even more important was the carefully constructed pattern of questioning, with repeated questions to gauge consistency and 'control questions' that were certainly right or wrong and served as benchmarks against which less predicable answers could be judged.

Polygraphs remain controversial. Their reliability and usefulness has been both defended and attacked, though in general lie-detector findings are not admissible evidence in court. We can argue as to whether they are more or less useful than an ancient Chinese technique: the mouth of the accused was filled with rice as the prosecutor presented evidence. One of the commonly accepted signs of stress is a dry mouth, since saliva production ceases. If the rice remained dry, the accused was considered guilty.

The Band-aid and the Tampon

Earle Dickson, Earl Haas

Do-it-yourself medical products sold in chemists and supermarkets provide ready solutions to conditions from a cut finger to menstruation. They make look simple, but someone invented them.

In 1921 Earle Dickson was a cotton buyer for Johnson and Johnson, which made bandages and medical adhesive tape among other things. Dickson's young wife Josephine liked to cook but was rather accident-prone, regularly cutting and burning her hands. Dressing the wounds was a two-stage process, with first a piece of gauze and then some tape to hold the gauze on place. Both had to be cut to size and dressing wounds on your own hands was a trial. The dressings often came off. Like every innovator, Dickson wondered it there was a better way.

The task proved easy. He cut a long piece of tape and placed small pieces of gauze at intervals down the middle. He covered it with fine cloth (crinoline) to keep it sterile. All his wife had to do was cut off a piece of the tape, pull off the crinoline and stick the dressing on. Dickson showed his invention to his boss, one of the Johnsons, who liked it and proposed marketing what would become J&J's most famous product. Dickson was promoted to Vice President. Early sales of the hand-made dressings were slow until

samples were given to the Boy Scouts to generate some publicity. Three years later, individually wrapped Band-aids were being made by machine. From 1939 they were sold sterilised. Vinyl replaced cloth tape in 1958, a year after the plant to manufacture them was moved to Dickson's home town.

Unlike the Band-aid, the tampon has had a long history. Ancient Egyptians and Greeks used softened papyrus or lint around a little piece of wood. Later they were made of paper, wool or cotton, but only one at a time. The twentieth-century approach in most things was mass production. Denver doctor Earle Haas, keen to find a replacement for the traditional 'rags' his wife and other menstruating women used, came up with a design and patented it in 1931. Basically a plug of compressed cotton sewn into a cotton bag, the tampon had an applicator for insertion, two cardboard tubes sliding one inside the other.

Haas was turned down several times by manufacturers, including Johnson and Johnson. So he sold his patent to a local entrepreneur, Gertrude Tendich, an ambitious woman recently migrated from Germany. She brought the Tampax tampon to the market in 1936, running up early batches on her home sewing machine.

Storing Food, Toasting Bread

Charles Strite, Earl Tupper

Tupperware and electric toasters, two of the many domestic innovations found in today's kitchens, have many differences, but some things in common. Both depended on certain crucial innovations, but also drew on wider trends in technology, the increasing availability of electricity in homes and the appearance of new plastic materials.

New Hampshire-born farm-boy Earl Tupper displayed an early flair for door-to-door salesmanship. However, the Depression forced his tree-surgery business into bankruptcy and the 30-ish Tupper found work briefly in the plastics division chemical firm DuPont. A year later he founded his own plastics company, supplying, among other things, gas masks to US troops during World War II.

At war's end, he decided to focus on plastic products for the home, a bold move since plastics were unfamiliar to householders. His subsequent success was built on three innovations. First, drawn from his DuPont experience, he transformed polyethylene slag—a cheap, black, evil-smelling by-product of petroleum refining—into a flexible, translucent material suitable for unbreakable cups and bowls. Second, he developed an air- and water-tight seal, similar to that on paint tins, for food-storage containers.

The last innovation was the 'Tupperware party', a gathering of potential customers in a private home. He did not invent this but he certainly exploited it to the full. It is, uniquely, still the only way to buy Tupperware. Tupper sold out in 1958 after a decade in the business, but the Tupperware age endures.

Even with the example of the incandescent light bulb (p. 137), electric toasters could not replace traditional forks over the fire until someone solved a crucial technical problem. The current-carrying wire in the toasting element needed to get hot enough for the toasting but not become so fragile that it would burn out. An engineer called Albert Marsh was the first to solve that, with his discovery in 1905 of an alloy of nickel and chromium called 'nichrome'.

Various models of electric toaster appeared, and sales rose sharply in the 1920s, following the patenting by Charles Strite of the 'pop-up' toaster, equipped with a spring-loaded timer to eject the toast when it was browned on both sides as required (determined by turning a dial), and before it burnt. It was an early example of a 'set and forget' household appliance. With a few refinements, mostly in styling, today's toasters closely resemble Strite's 1926 Toastmaster.

The Movies Learn To Talk

The Legacy of *The Jazz Singer*

Hollywood feigned indifference when the technology for 'talkies'—movies with speech synchronised to the actors' lips—was convincingly demonstrated. Studios had huge inventories of silent films and actors on long-term contracts, many with heavy foreign accents. The cost of converting theatres to sound would be colossal. But in their hearts, the studios knew that making movies talk would greatly increase their appeal, making them mainstream entertainment. Progress might be delayed but it could not be denied. In October 1927 Warner Brothers released *The Jazz Singer* with Al Jolson, the first feature-length talking picture, though not the first movie with sound. There was no going back.

The move to talkies saw a battle between competing technologies, sound-on-disc and sound-on-film. The first was pioneered by Thomas Edison, inventor of the phonograph (p. 135), but he later decided it had no future. Western Electric and Bell Telephone used the new electronic amplifiers (p. 176) to create the Vitaphone ('sound-on-disc') process, used in *The Jazz Singer*. Vitaphone had problems.

It was notoriously difficult to synchronise the recorded sound with the film. The record could jump a groove, the film might have been damaged and some of it cut out. Either event could destroy synchronisation, needing swift action by the projectionist. Frequent playing wore the discs; they had to be replaced every 20 showings.

Initially, discs provided better sound quality, but sound-on-film quickly improved, and the other advantages soon won out. The soundtrack was recorded directly onto the film beside the picture so synchronisation was automatic. To reproduce the sound, a light shone through the soundtrack, which varied in opaqueness in line with the sound, and onto the newly invented 'photocell'. This produced a varying current that went through amplifiers to speakers.

Victory for the new technology was swift, beginning with the 'all talking, all singing, all dancing' *Broadway Melody* from MGM in 1929, which won Best Film at the Oscars. Warner Brothers stopped making new Vitaphone movies in 1930. The cost of converting theatres to the new technology meant alternative sound-on-disc copies were needed until 1937, but the technology was doomed.

Something else was doomed—the habit of talking while the film was screening. This was frowned upon in the new talkies era. As one commentator put it, 'talking audiences for silent films were replaced by silent audiences for talking films'.

The First Antibiotics

Alexander Fleming, Howard Florey, Ernst Chain

Since the discovery in the nineteenth century that infections are caused by microbes, medical researchers had sought better ways to kill microbes and control often life-threatening infections. Scottish physician Alexander Fleming, working at St Mary's Hospital in London in the 1920s, was keen to find better ways to control infection in wounds, including surgical ones. He knew from experience in the trenches of World War I that infection could kill more soldiers than bullets or shells.

Fleming's famous discovery (it was not yet an invention) occurred mostly by chance. Returning to his laboratory from a holiday in 1928, he examined some unwashed glass dishes on which he had been growing the sometimes-deadly bacteria *Staphylococcus aureus* ('golden staph'). He noted that a blue–green mould was growing on one dish, and that the growth medium on the dish was free of bacteria around each cluster of mould. Was something seeping from the mould stopping the bacteria growing?

The mould had apparently come from soil in the garden, blown in through an open window. Fleming identified it as *Penicillium notatum,* so he called the yet-to-be-identified bacteria-killer penicillin. There was scant evidence of its ultimate value. Fleming reported that it was unstable and difficult to purify. Fleming was a doctor, not a chemist and so not the man for the job. After a few years, the work petered out.

Fleming had made the key discovery; credit for the innovation lies elsewhere. A decade later, German-born chemist Ernst Chain, working in Oxford with Australian Howard Florey, came across a report of Fleming's discovery. The pair decided to invent a way to mass-produce penicillin; that goal required significant innovation. The results were startling. Penicillin cured bacterial infections in mice, then in humans, and by the end of the war was saving the lives of seriously wounded soldiers. Soon half a tonne of penicillin was being produced every month.

Available to the general public once the war ended, penicillin was the first of the 'antibiotics'; different moulds would later produce other drugs such as streptomycin. Together with the 'sulpha' drugs, also quite new, antibiotics proved effective against many infections. Ominously, strains of bacteria not killed by these 'wonder drugs' began to emerge within a few years. Such drug resistance keeps ingenious chemists busy devising new antibiotics to this day. Often the bacteria seem to be winning.

Reinventing Ancient Toys

Pedro Flores, Donald Duncan, Richard Kneer, Arthur Melin

The hula hoop and the yo-yo are ancient entertainments, if not as old as the doll. In the twentieth century, they were rediscovered and reengineered, and marketed so well that they became 'crazes', enjoying recurring periods of intense popularity.

The Hula Hoop

Many cultures played with hoops of wood, bamboo or metal for millennia, but the modern resurgence began in the 1950s. Australian school children twirled cane hoops around their waists in exercise classes. Hearing the news, Americans Richard Kneer and Arthur 'Spud' Melin, who sold the Wham-O slingshot, grabbed the opportunity. Their hoops of brightly coloured plastic, often with ball bearings, bells or other noise-makers inserted, sold 25 million in two months, and 100 million during 1958. It was the biggest toy fad in American marketing history.

The 'hula hoop' (so called because the hip movement needed to keep the thing in motion resembled Hawaiian dances) was cheap and inclusive. People of any age could use one and devise their own style. Hoop-spinning contests were held at local fairs. An accomplished performer once spun 82 hoops simultaneously. That made the *Guinness Book of Records*.

The Yo-yo

Stone versions of the yo-yo date back at least 3000 years. Many cultures used the toys, made from wood or bone, and they were popular, under another name, among eighteenth-century French nobility. The man responsible for starting the first modern craze appears to have been Pedro Flores. Moving to the USA from the Philippines, he started manufacturing yo-yos in 1928; the name is Filipino. In a significant technical advancement, he looped the string around the axle, rather than affixing it. This let the yo-yo spin at the end of the string, making possible feats like 'walking the dog' and 'around the world'. By 1929 yo-yos were selling by the million, and hobbyist magazines carried articles on 'how to make your own'.

The yo-yo did not make Flores rich. He had already sold out for $25 000 to Donald Duncan, who later claimed 85 per cent of the market. Duncan had a genius for publicity, sponsoring contests and getting movie and sports stars to pose yo-yo in hand. In the 1960s he employed dozens of professional demonstrators across the country; his factory made 3000 yo-yos an hour. He was too successful; 'yo-yo' was declared a generic name. Loss of trademark protection bankrupted Donald, though he kept himself going marketing parking meters (p. 203).

Fast-frozen with Flavour

Clarence Birdseye

The English philosopher Francis Bacon reputedly died in 1626 from a chill caught while stuffing a chicken with snow as an experiment in food preservation. But not until the 1920s did anyone devise a commercially viable food-freezing process that retained flavour and texture. His surname is still seen in many supermarket freezer cabinets: Clarence Birdseye.

Brooklyn-born Birdseye always loved the outdoors. He hunted and fished, paying his way through college by catching rats for sale to a laboratory, and spending five years in Labrador trapping foxes for fur. There he observed the traditional Eskimo method for preserving fish. Thrown onto the ice, fish froze almost instantly in the frigid air. Birdseye noted that the freezing took place so quickly that only very small ice crystals formed; tissues were less damaged and when thawed the fish tasted much better than if frozen more slowly. Living in a remote location, Birdseye froze vegetables like cabbage to give his family fresh food.

Back home, he applied what he had seen in Labrador to large-scale food preservation. In 1923, using a fan, tubs of brine and cakes of ice (total cost $7) he experimented in his kitchen with quickly freezing rabbit meat and fish fillets, packed into waxed boxes ready for defrosting and cooking without further preparation. His innovation combined convenience with flavour. Turning that process into a viable business took time and Birdseye went broke once. In March 1930 a range of frozen meats, fish, vegetables and fruits went on sale in 18 stores in Springfield, Massachusetts. Sales were slow initially: prices were high and store displays unattractive.

By the mid-1930s the idea was catching on. Birdseye leased freezer cabinets to retailers who could not afford to buy them during the Depression. By 1944 he was leasing insulated railroad cars to send frozen food around the country. The amount of space in the new-fangled home refrigerators devoted to frozen foods grew as householders bought more. Today frozen food is a major industry, relying on blasts of extremely cold air.

Birdseye's inventiveness went beyond frozen foods, securing him 300 patents. He was proudest of an improved process to convert sugar cane waste into paper pulp in 12 minutes instead of nine hours. He attributed his success not to outstanding intelligence or sales skill but to intense curiosity combined with a willingness to take risks.

Fun in the Arcade

Raymond Maloney, Harry Mabs

During the dark days of the Great Depression, the unemployed flocked to the 'penny arcade' in search of cheap, time-filling entertainment. Popular among the attractions was the 'pinball' machine. A metal ball was fired across a sloping, glass-shrouded board, studded with metal pins and equipped with a few holes into which the ball could drop as it rattled among the pins. A ball in the right hole could win the player a useful prize.

The pinball machine, based on a nineteenth-century entertainment known as a 'bagatelle', was coin-operated, as were the earliest jukeboxes based on Edison cylinders (p. 157). The first such machine was built in 1931 by Raymond Maloney from the Bally company, later better known as a maker of slot or poker machines. From the start, makers were constantly innovating, as players were always in search of something new to hold their attention. Once machines were electrified, bells, buzzers and flashing lights added to the excitement. Machines were soon equipped with a 'tilt mechanism'— sensors detected if a player tried to alter the path of the ball (and their chances of winnings) by lifting, shaking or thumping the machine. Other versions saw the pins and holes replaced by spring-loaded bumpers. Hitting the bumpers scored points, which were automatically totalled.

Early pinball machines were more a matter of chance than of skill. Indeed in 1942, the colourful Mayor of New York, Fiorello LaGuardia banned them on the grounds that the games actually constituted gambling. The ban lasted till 1976. In response to the ban, or perhaps just to provide something new, Harry Mabs from the Gottlieb company added electrically powered 'flippers' to the new Humpty Dumpty pinball machine in 1947. These were operated by the player to keep the ball in play, prolonging the enjoyment and adding a substantial measure of skill to the game. Mayor LaGuardia should have been satisfied.

Innovation continued, with new layouts, more exciting lights and sounds, machines that played several balls at once, digital readouts, moving displays using LEDs or video screens, workings based on integrated circuits rather than electromechanics, and machines that talked. But pinball machines and similar pastimes could not keep pace with the growth of purely electronic 'video games', in the arcade or at home, and the number of manufacturers in the USA is now down to one. Interestingly, one popular video game is a computer simulation of pinball.

Beyond the Light Microscope

1931

Ernst Ruska, Max Kroll

The invention of the microscope in the late sixteenth century (p. 21) transformed biology and other early sciences. The microscope made it possible to view objects too small to see with the unaided eye—previously unknown or even unsuspected objects like the bacteria and yeasts that were later found to cause disease and fermentation. But microscopes using light were limited to magnifying 500 or 1000 times, and could not image objects much less than a micron (a thousandth of a millimetre) in size. Those limits had been reached by the 1930s.

In the hunt to see even finer detail, inventors began using beams of electrons, the charged particles that make up an electric current or the cathode rays in discharge tubes (p. 161). Since a beam of electrons could be focused by a magnetic field, just as light is focused by an ordinary lens, an 'electron microscope' looked feasible. And as the radiation associated with the electron beam has a much smaller wavelength than visible light, an electron microscope should be able to magnify very much more, perhaps revealing objects nanometres (millionths of a millimetre) in size rather than mere microns.

The first primitive models made by Germans Ernst Ruska and Max Kroll in 1931 could manage only 400 times magnification, but the technology progressed rapidly. By the mid-1930s, 10 000 times magnification was possible, reaching 200 000 times by the 1960s. New discoveries poured from laboratories using these instruments: the delicate internal architecture of plant and animal cells, images of disease-causing viruses (known to exist but never before seen) and the even more remarkable 'bacteriophages', resembling hypodermic needles on legs and able to infect bacteria. For his invention, Ruska shared the 1986 Nobel Prize for Physics.

Like light microscopes, the first electron microscopes shone their radiation through the specimens, producing a cross-section, referred to as 'Transmission EMs'. In the 1940s these were joined by 'Scanning EMs' (SEMs). These ran the electron beam back and forwards across the object, collecting the electrons that bounced off and forming a remarkable three-dimensional image. This realism is evident even when the images are not highly magnified. An SEM of a tiny ant, for example, reveals a huge beast seemingly ready to spring from the page; a human hair looks like a tree trunk. Under extremely high power, an SEM can show individual atoms, looking like solid balls of matter.

A New Window on the Universe

Karl Jansky, Grote Reber

It is a major achievement to make the previously unseen visible, and to do it by accident. Yet that is how Karl Jansky, an engineer at the Bell Telephone Company in New Jersey, stumbled on the 'radio telescope'. The company's radio links across the Atlantic and Pacific were plagued by static, which crowded in on the traffic, making messages hard to read. Jansky was asked to find the source of the interference.

Using a large radio antenna on a rotating platform, Jansky soon identified thunderstorms, other transmitters, noisy electrical equipment in factories, and aircraft as obvious sources of static. Yet even discounting those, a faint radio hiss continued to arrive from nowhere in particular.

Jansky ultimately found that the radio noise was strongest when the Milky Way, an immense collection of stars made faint by distance, was visible overhead, particularly its thickest, widest part. Here was evidence that radio waves as well as light rays come from the heavens far beyond the Earth, and that a radio receiver pointed toward space might reveal objects and events that could not be seen.

Understandably, that was of no interest to the Bell Telephone Company. There was nothing they could do about an 'extraterrestrial' source; they closed the project down and moved Jansky onto other duties. Professional astronomers also took little interest and frowned on the suggestion that radio waves might come from the Milky Way. What could cause them? The only follow-up was by a radio amateur called Grote Reber, who built the first radio telescope 'dish' in his backyard and drew the first maps of the strength of radio noise across the sky. Again the astronomy community ignored the work.

From this discovery did in time come the whole new science of radio astronomy. Observations made with radar sets during the world war that followed stimulated much interest, and in the post-war world astronomers began to build big instruments to explore the 'radio universe'; helped by burgeoning radio technology. Fifty years later, hundreds of radio telescopes dot our planet, many of them linked together to improve resolution and sensitivity.

The same path is being followed with other forms of invisible radiation from the cosmos. Highly advanced instruments on the ground and in space collect and analyse infrared and ultraviolet light, microwaves, X-rays and gamma rays to build a more complete understanding of what the universe is saying to us.

A Life for Radio

Edwin Armstrong

When American radio pioneer Edwin Armstrong jumped from his thirteenth-story New York apartment in 1954, bankrupt and distraught, he ended a brilliant career of innovation that still brings everyday benefits. In particular, his promotion of frequency modulation (FM) broadcasting ultimately transformed the radio industry. His genius had been recognised early. In 1917 he had been the first recipient of the Medal of Honour of the Institution of Radio Engineers.

Ironically, FM drove him to suicide. Armstrong had wanted to improve the quality of radio reception, particularly for music. Since the days of Lee de Forest (p. 176), radio used amplitude modulation (AM), the amplitude (wave height) of the radio signal varying in line with the sound signal being impressed upon it. Unfortunately, electrical sparks (such as lightning), passing tramcars or faulty motors generated similar waves, causing troublesome static or interference. Armstrong devised another approach, slightly varying the frequency (the number of cycles per second) of the radio signal to code the words or music. The result was clearer, static-free sound.

Armstrong, then working for the Radio Corporation of America (RCA), patented FM in 1933. By the early 1940s a number of FM stations (the Yankee Network) were operating in the New England region, using a radio band allocated by government regulation. FM threatened the profitability of the hundreds of AM radio stations then on air. RCA owned many of these and fought back by getting the government to shift the FM band, ostensibly to make room for the coming of television, but really forcing FM into a less suitable frequency range and undermining the stations.

RCA also claimed patent rights over FM, since Armstrong had been working for RCA when he invented it. Armstrong lost the subsequent court battle. Unable to claim royalties for FM sets then in use, he was soon ruined; suicide quickly followed. His widow continued the fight, finally prevailing in 1967. After decades of technical development, FM also triumphed.

Armstrong was no stranger to patent battles. In the 1920s, he had struggled against de Forest and his backers over patent rights to a new sort of radio receiver using the 'regenerative principle'. Most of its internal operations took place at one frequency, so it was simpler to tune from one station to another. The decade-long lawsuit was ultimately settled in de Forest's favour, though not everyone accepts the Supreme Court's judgement, even today.

Electronics Makes Music

Laurens Hammond, Léon Theremin

Electronic music did not begin with the now-ubiquitous synthesiser or keyboard, an application of digital information technology. Following the invention of radio valves to amplify and oscillate (p. 176), bright minds found ways to use them in musical instruments, rather than relying on vibrating strings or air columns. Two such devices, invented at much same time, were the Hammond organ and the Theremin. The Hammond organ, patented in 1934, is more familiar, with a keyboard and a recognisable 'theatre organ' sound, much associated with American gospel music. Its method of making music was altogether new.

Soon after service in the trenches of World War I, American engineer Laurens Hammond invented the 'synchronous electric motor', which kept in precise step with the 60 cycles-per-second alternating current used to distribute electricity. He first used this motor in a clock that did not tick.

He soon realised that the motor's unwavering rotation could generate precisely-pitched musical notes by spinning wheels with varying numbers of teeth through electromagnetic fields. So the Hammond organ was an 'electromechanical' device, like some early computers. Loudspeakers using a revolving disc produced the unique trembling sound. Hammond organs could be a one-person band, with attachments to recreate the sound of a bass instrument or of percussion. They have recently undergone a revival among devotees of funk music, but the massive old-style Hammond organs are no longer made.

The theremin, created by Russian physicist and cellist Léon Theremin, was very different. With no keyboard, its notes were controlled entirely by the movement of the player's hands, effectively part of the electronic circuits. One hand determined the pitch of the rich sound, the other its volume. Theremin visited the USA in 1927 to demonstrate his instrument. It generated much interest, largely through the mesmerising skill of his young protégé Clara Rockmore. Her advanced sense of pitch and precise control of movement allowed her to master an instrument unlike any other.

Despite its ingenuity, and its place in the development of electronic music, the theremin was not a commercial success. The electronics manufacturer RCA put it on the market but sold only a few hundred. Yet it is likely you have heard its unforgettable, eerie tones in the soundtracks of classic movies like *Spellbound, The Lost Weekend* and *The Day the Earth Stood Still*.

Paying to Park

Carl McGee

Traffic and cities have long gone together, especially once the horseless carriage replaced the horse and buggy (p. 182). As motorcars began to throng city streets, there was soon nowhere to park.

Carl McGee, newspaperman and civic leader, was chair of the Traffic Committee of the Chamber of Commerce in Oklahoma City in the USA. With his colleagues he faced the challenge of the 'all-day parkers'. Those who worked in the downtown area left their cars parked alongside the footpath right through working hours. This forced shoppers to park a long distance away, which did not impress retailers. Time limits were imposed and traffic police chalked tyres to keep track of time, handing out tickets, but the problem did not go away.

McGee decided that he needed an inexpensive machine that would be stationed adjacent to each kerbside space, and which could record how long each car stayed parked. Helped by staff and students at Oklahoma State University, he devised a machine nicknamed the Black Maria, and in 1933 went looking for a manufacturer. The early machines were driven by clockwork, so the tender went to a firm that made clockwork fuses to set off nitroglycerine in oil wells.

The first 175 parking meters were installed in 1935, and proved so successful that they were soon installed all over town. The rest of the world followed, inspired by the threefold success of McGee's innovation. The machines straightened out the parking problem, and the ten-cents-an-hour fee for legitimate parking, together with a $20 fine on those who stayed too long, brought extra revenue to the city. The values of downtown properties went steeply up as the parking nightmare disappeared. Indeed, meters quickly became so popular with retailers that everybody wanted one outside their store. But not everyone welcomed the meters. Some motorists who objected to paying for something they thought should be free tried to disable the meters, a not-uncommon fate for innovations.

Parking meters today work and even look much like McGee's, but innovation has continued: initially two meters were put on the one pole to reduce clutter ; more recently meters were introduced that can print out tickets for four to six spaces, the motorists then being required to display the ticket behind their windscreen. This has cut costs and boosted revenue.

Movies in Colour

Herbert Kalmus

Despite early triumphs in photo-graphy (p. 85) and motion pictures (p. 158), capturing realistic colour proved difficult. We knew the science behind colour perception well enough. In the 1860s Scottish genius James Maxwell had created coloured images by taking three separate pictures through red, green and blue filters and then projecting them top of each other through the same filters. Though complex, 'additive' technology delivered the earliest colour prints.

It was also used in the first successful process for colour movies. 'Technicolour' was invented by Herbert Kalmus and his colleagues early in the twentieth century (the 'Tech' in Technicolour recalled their alma mater, the Massachusetts Institute of Technology). Light entering the camera was split into separate beams, passing through coloured filters (initially two, red and green, but a third blue filter was added later) and then onto two (later three) strips of monochrome film joined together and running side by side.

In projection, the developed strips of film ran through the projector side by side, separate light beams passing through filters and blending into the coloured moving image on the screen. Much depended on the skill of the projectionist.

Kalmus was on surer ground using the black and white originals to print three coloured images one above the other in a single film and projecting one beam of light through the lot. The effect was brilliant. Walt Disney adopted Technicolour very early and that helped explain the popularity of his animated films in the 1930s. Since the original prints were held in black and white, new colour prints could be made at any time, making Technicolour effectively immune from fading.

The system was costly and cumbersome, and the hunt was on for a comparable single-film process. This involved passing the light through three transparent light-sensitive layers one after the other, each recording a primary colour. When the three developed layers were viewed together, the result was very effective. Eastman Kodak put the first still film (for 'snaps') using this process onto the market in 1935 (the same year *Becky Sharp*, the first movie using three-colour Technicolour, came out), and marketed a movie version called Eastmancolour in 1952.

The technology continued to advance, and manufacturing colour films became a remarkable achievement, requiring a dozen separate chemical-packed layers, precisely laid to make a strip less than a millimetre thick. Yet, however clever, it is all old-hat now that we have digital cameras (p. 279).

Seeing by Radio

Robert Watson-Watt

An early discovery about radio waves was that objects reflect them, just as they do ordinary light. Sharp minds, such as radio pioneer Reginald Fessenden (p. 177), argued that such reflections could reveal the presence of things hidden by darkness, fog or distance. This might stop ships colliding with each other or with icebergs, for example.

Lack of real interest, and the limited technology of the day, saw the idea languish until the 1920s, when its potential military value emerged. Could hostile aircraft or warships be detected when they were still a long way off, giving time to organise defence? Research began in several countries, but again there was no sense of urgency.

In the mid-1930s British authorities suddenly realised the potential for devastating air attack by a rearming Germany. 'The bomber will always get through', cried the pessimists. As a result, interest in radio-location revived in 1935, led by Scottish-born engineer Robert Watson-Watt, who can claim to be the inventor of 'radar', as it was later called (for RAdio Detection And Ranging), at least in Britain.

A beam of radio waves was initially seen as a potential 'death ray', disabling the pilot of an attacking aircraft by heating his blood. That quickly proved impractical, but the radio reflections from his plane could certainly be detected, and systems were soon developed to measure the distance, height, direction and numbers of attacking forces. The first stations were operational in 1936, and by the time war began in late 1939, a defensive chain was in place around the southeast coasts of Britain.

Was radar 'the weapon that won the war', rather than, say, the atomic bomb (p. 216) as many would claim? Certainly radar made a crucial difference in many significant theatres of the conflict, helping the RAF triumph in the Battle of Britain in 1940, blunting the impact of German night-bombing during 'the Blitz', winning the Battle of the Atlantic against the U-boats, guiding bombing raids into occupied Europe Other countries developed it too, but Britain used it best.

As with radio during World War I (p. 189) military demands drove the technology of radar rapidly forward, incidentally readying it for civilian use when peace came. One of the many outcomes was the mastery of microwaves, which would find a multitude of uses (p. 219).

Television Goes on Air

Vladimir Zworykin, Philo Farnsworth

Science and technology often take wrong turnings, and end up exploring blind alleys. So it was with embryonic television. The notion of using Karl Braun's cathode ray tube (p. 161) to capture and display images, proposed in 1908 by the Scots visionary Alan Campbell-Swinton (p. 179), was sidelined for several decades by a mechanical system of scanning first proposed by German Paul Nipkow. In the 1920s two inventors took this up: American Charles Jenkins and Scot John Logie Baird, the man often (and incorrectly) credited with single-handedly inventing TV.

Both of these men developed TV systems based on scanning discs, and had many followers. In the receiver, the image was reconstructed by using the incoming signal to change the brightness of a lamp that shone through a second scanning disc onto a screen. Dot by dot the image would appear, with the persistence of vision of the eye blending the trail of bright spots into a complete picture. Baird's system was publicly trialled in Britain in the 1930s. But the pictures were crude and fuzzy (Baird's system for example had only 30 scanning lines to the frame). Scanning discs were soon abandoned in favour of the electronic systems prophesied by Campbell-Swinton.

Others, such as Russian Boris Rosing in 1907, had tried using Braun's invention to make TV work, though the scanning in his transmitter was done with a Nipkow disc. Rosing did manage to transmit some crude geometrical shapes, but pictures with various shades of grey were beyond him and without the aid of electronic amplifiers, then only just being invented (p. 176), the images were very faint.

Rosing's student, Vladimir Zworykin had migrated to the USA in 1919 and joined the staff of Westinghouse; it was he who did most to realise Campbell-Swinton's vision. In 1923 he patented an electronic camera based on the cathode ray tube. Initially it was no match for the ubiquitous scanning discs, but by 1928 the 'iconoscope' was ready for use and for a decade of legal battles over patent rights to the invention.

At least one other name must be mentioned—that of Philo Farnsworth. He made so many contributions that many regard him as the real inventor of television. Professor Farnsworth in the TV series 'Futurama' carries his name. He devised his first system at the age of 15 as a Utah farm boy, inspired by taking part in his first telephone conversation. Through the 1920s he pioneered electronic systems in parallel with Zworykin, in particular building his 'image dissector'. Modern video cameras are based on a combination of Zworykin's and Farnsworth's ideas.

It was reputedly Farnsworth who convinced John Logie Baird that his mechanical systems could not compete with electronic ones. By 1935 he had built a complete television system, used a year later to transmit entertainment programs. Farnsworth struggled with depression late in life, and until his death in relative obscurity in 1971 was preoccupied with trying to control nuclear fusion at a power source.

The first regular television broadcasts went to air in 1936, less than two decades behind radio (p. 189). Transmissions by the British Broadcasting Corporation (BBC) originated from Alexandra Palace in London, using both electronic and mechanical scanning for a time, and televising the coronation of George VI. The outbreak of war in 1939 closed TV stations down, though tremendous technical advances driven by war needs, such as those of radar (p. 205), made television a far more powerful medium when large-scale transmission resumed.

In the USA, television quickly came under the control of commercial radio interests. This move was led by David Sarnoff's (p. 189) Columbia Broadcasting System (CBS), which had paid for much of Zworykin's research, and the National Broadcasting Company (NBC). They saw television as supplementing the mix of entertainment and information that that was working so well on radio. First to air was NBC in 1939, telecasting President Roosevelt opening the World's Fair in New York.

The public seemed slow to accept the new wonder. After two years of BBC telecasts, only 3000 sets were in use, and only 500 were bought in the first six months after NBC went to air. The high cost of the equipment and the crudeness of the images were major disincentives, and programming was rudimentary compared with the sophistication of radio. Modern TV is a world away from its 1930s beginnings (p. 234).

Sales picked up after the war as public acceptance grew, along with the quality of equipment and programs. The 7000 sets in the USA in 1945, with only five stations broadcasting, grew to 10 million sets and 100 stations in 1950. Serious drama, news and public affairs filled the schedules until the mid-1950s and the rise of game shows and 'sitcoms'. Television spread quickly throughout the industrialised world, reaching Australia, for instance, in 1956 in time to broadcast the Olympic Games from Melbourne.

Jet Engines Replace Propellers

Frank Whittle, Hans von Ohain

The jet engine is a great example of independent invention, an idea 'whose time had come'. Two military aviation engineers, Frank Whittle in Britain and Hans von Ohain in Germany, unknown to each other, developed almost the same ideas through the 1930s.

Since the Wright Brothers (p. 172), aircraft had been driven by propellers, that technology peaking in fighter planes like the German Messerschmitt and the British Spitfire and Hurricane. But the limitations were already obvious: no propeller-driven plane, it was believed, could ever exceed the speed of sound. A new approach was needed.

Whittle and von Ohain had both designed what was essentially an air-breathing rocket. A traditional rocket had in its mix of chemical fuels one that supplied oxygen to make the rest burn (salpetre in the case of gunpowder); the 'jet engine' would take its oxygen from the air it flew through, compressing and heating the air with a turbine (p. 142), spraying in liquid fuel and igniting the mixture. Hot high-speed gases injected from the rear would cause the engine, and the aircraft that carried it, to move forward, as a rocket did. This would make more efficient use of the fuel, generating more power and more speed, with fewer moving parts, though there would be major engineering challenges.

Frank Whittle patented his jet engine in 1932, though many doubted the idea had real merit. He struggled for funds and recognition, battling bureaucrats and potential industry partners, but the outbreak of war in 1939 spurred action and his E-28 prototype engine flew successfully in 1941. The rewards came later, a knighthood, promotion to Air Commodore, a substantial ex-gratia payment from the government and the knowledge that, by war's end, Meteor jets using his engines were destroying German V1 rocket bombs directed against London.

Like Whittle, Hans von Ohain had been only 22 years old when he first conceived an aircraft engine without a propeller. He patented his design in 1934, several years after Whittle, but with financial backing from leading aircraft designer Ernst Heinkel, he put his ideas to work much more quickly. His jet engine ran successfully on a test bench in 1937 and powered the Heinkel He178 in the air in August 1939, two years ahead of Whittle, and just a few weeks before war broke out. Germany maintained its lead when the Messerschmitt Me262 became operational before the Meteor.

Machines To Replace Organs

William Kolff, John Gibbon

Some people have defective kidneys, unable to adequately cleanse their blood of wastes, so threatening their health and even life. Others must manage without their hearts for a time, so they can be operated on. The triumphs of biomedical engineering include machines to mimic those organs—the dialysis machine, which supports ailing kidneys, and the 'heart–lung' machine, able to do the work of both for a time.

Dutch doctor William Kolff created the 'kidney machine'. In 1939, he filled 50 centimetres of cellophane tubing (actually artificial sausage skin) with blood containing a lot of urea, a waste from the breakdown of proteins. He then shook the sealed tube vigorously in saline (salt) solution. Much of the urea passed through the walls of the tube into the saline, leaving the blood greatly purified. In 1943, working under difficulties during the Nazi occupation, Kolff mechanised the process and made it continuous. A long tube around a drum rotating in saline drew blood from one of the patients' arteries and returned it to a vein. During the war, he treated 15 kidney patients in this way, thought only one survived long term.

Migrating to the USA when the war ended, Kolff made many improvements; dialysis machines became smaller, cheaper and simpler to use. But the real impact did not come until the 1960s, when techniques were developed to enable patients be connected and disconnected quickly and without the need for surgery.

The 'cardiopulmonary bypass machine' (aka the heart–lung machine) was pioneered by American cardiac surgeon William Gibbon in 1937, working at the same time as Kolff and stimulated by the death of a heart patient some years earlier. Gibbon argued that bypassing the heart would allow time for careful heart surgery, though he got little support for the idea. His experimental devices pumped blood around the body and also oxygenated it, so duplicating the operation of the lungs. Doing both greatly simplified the procedure.

After early experiments on cats and dogs, Gibbon gained financial support from IBM Chairman Thomas Watson, improving the machine (such as reducing damage done to blood passing through) to the point where human use was safe. In the first ever 'open-heart' operation in 1953, he closed a hole in the heart of 18-year-old Cecilia Bovalek. During the operation, the machine replaced her own heart and lungs for 26 minutes. In 1967 South African surgeon Christian Barnard used a similar machine during the first heart transplant.

The First Photocopy

Chester Carlson

Modern technology was predicted to bring the 'paperless office', yet today's workplaces use more paper than ever. Our paper copiers have unparalleled productivity. Yet for generations the only copying methods were carbon paper, useful only when a document was first prepared, and the photostat, really a form of photography, which needed chemical development and took time to dry.

Seattle-born Chester Carlson had degrees in physics and law. He headed the patent department of an electronics firm, but it was the Depression and times were tough. He set out more or less deliberately to invent what we now call the photocopier, partly to earn some more money, partly to make extra copies of patent applications. Months of library research, his own background in physics and years of experimenting (in his kitchen until his wife complained) ultimately produced success. Carlson's experience symbolised Edison's famous dictum about '1 per cent inspiration, 99 per cent perspiration'.

If a light image of a document was made on an electrically charged plate (initially zinc covered with sulphur), the plate remained charged only where the image was dark. Carlson then covered the plate with black lycopodium powder, made from club moss spores. The powder stuck where the plate was charged (the dark parts of the image) but not elsewhere. Loose powder was blown away. Finally Carlson pressed a piece of heated waxed paper onto the plate. When he peeled it off, the adhering powder came away as a copy of the original.

Carson secured a patent in 1938 but no one else thought the invention had a future. Over the next five years IBM, Kodak, General Electric and RCA, among others, declined to invest. Finally the non-profit Battelle Memorial Institute took an interest. Researchers there made some major improvements, using selenium rather than sulphur to coat the plate, and mixed fragments of iron, ammonium chloride and plastic as the powder.

By 1948 crude, primitive photocopiers were being marketed by Haloid, a small photographic company that took a major risk on the untried technology. Early machines needed 14 steps and 45 seconds to make a single copy; they hardly sold at all. Another 10 years' work produced the first automatic photocopiers, and those sold so well that Haloid earned $60 million in sales in 1960. The patient Carson got his reward, most of which he gave to charity. By then, Haloid had renamed itself Xerox Corporation, and the process 'xerography', from a Greek word meaning 'dry'.

DDT: A Wonder Chemical?

Paul Müller

Our ambivalence about the insecticide DDT pits its unprecedented success in controlling serious diseases carried by insects against its proven harm to much wildlife, as set out in biologist Rachel Carson's highly influential 1962 book *Silent Spring*. Dichloro-diphenyl-trichloro-ethane was first synthesised in 1874 by Othmar Zeilder, a Viennese pharmacist, but he did not realise its properties. When rediscovered in 1938 by Swiss chemist Paul Müller, working for the Geigy company, its lethal impact was soon recognised. Ready victims included mosquitoes, fleas and other insects carrying malaria, yellow fever, typhus, bubonic plague, river blindness and elephantiasis, as well as agricultural pests such as potato beetles, cotton boll worms and coddling moths.

DDT was patented in 1940 and reached the market as a dust in 1941. Its use was greatly stimulated by the outbreak of the Pacific War, and the need to control pests in areas of conflict, especially in the tropics. It was routinely dusted on soldiers to control lice that spread typhus. The results of its later widespread use are striking. Spraying with DDT virtually eliminated malaria from 20 countries where it had been endemic. Death rates in India, for example, fell from one million a year to less than 5000, extending average lifespan from 32 to 47 years. The incidence of malaria in the USA dropped from 250 000 a year in the 1930s to less than 10 a year 20 years later. For his invention Müller was awarded the 1948 Nobel Prize for Medicine.

DDT appeared to be an ideal insecticide: it was inexpensive to make, long lasting in effect (so not needing frequent reapplication) and apparently not harmful to plants and animals. It seemed to deserve the 'wonder chemical' tag. However Carson's book, highlighting previously unsuspected damage to bird reproduction (hence the potential for a 'silent spring'), together with evidence from laboratory rats and mice, saw it banned in a number of countries and damned as one of the worst environmental pollutants. It remains in use in the developing world, where nothing matches its cost-effectiveness, particularly against malaria.

DDT dissolves readily in fat and so builds up in human body tissues, but the threat posed remains uncertain, unlike the millions of lives undoubtedly saved. The shortcomings of DDT, which included the appearance of insects resistant to its effects, have driven sophisticated research into alternatives over the 60-plus years since its first use.

The Psychedelic Sixties

Albert Hoffmann

The 1960s was the era of 'flower power', the anti-Vietnam War protests and the emergence of the Beatles and the Rolling Stones. It also saw to the rise to popularity of LSD or 'acid', a mind-altering chemical that spawned the cultural phenomena called 'psychedelia'. The name most associated at the time with LSD use was Timothy Leary, a Harvard University psychiatrist who experimented extensively on himself with the drug; but the man who first discovered, even invented, LSD was Swiss researcher Albert Hoffmann, who did not at first know what he had found.

During the 1930s Hoffmann worked for the pharmaceutical company Sandoz. He was making a variety of chemical compounds from lysergic acid, which other chemists had extracted from the ergot fungus that grows on rye and other grasses. Outbreaks of ergot poisoning, usually from contaminated bread, had occurred throughout history, with a variety of bizarre effects. Lysergic acid therefore seemed an interesting subject of study.

The twenty-fifth compound Hoffmann made that year (1938) was lysergic acid diethylamide, which he dubbed LSD-25. Not finding any significant reaction to it from his laboratory animals, he put it aside. Not till 1943 did he come back to it, and apparently absorbed a tiny dose through the fingertips of his ungloved hands while making some more. He quickly became 'semi-intoxicated', in his own words, his imagination strongly stimulated. With his eyes closed he saw 'an uninterrupted stream of fantastic pictures, extraordinary shapes with intense kaleidoscopic colours', and had to go home and lie down until the effect wore off.

Wanting to know more, he deliberately swallowed 35 millionths of a gram three days later. The dose was minute, but the effects occurred again, just as vividly. LSD would prove to be one of the most powerful hallucinogenic drugs ever developed, yet it left him 'without a hangover' the next morning.

Hoffmann believed that the unique mind-affecting properties of LSD would find use in neurology and psychiatry, and were worthy of intense study. In time, that gave rise to 'psychedelic therapy' as an alternative to psychotherapy. The drug successfully treated long-term alcoholics, and the CIA reportedly took an interest, on the grounds that enemy nations where ergot was common might have found strategic uses for LSD. But the drug fell from official favour and its sale was declared illegal in the USA in 1968. It may or may not be relevant, but Albert Hoffmann lived to be 100.

New Ways to Write (and to Correct)

The Biro Brothers, Bette Graham

During World War II, the British Air Force was looking for a new type of pen that did not leak at high altitude, unlike fountain pens. Such a pen already existed, patented in 1938 by Hungarian journalist Laszlo Biro and his brother Georg, but not yet attracting much attention. Biro had noted that ink used to print newspapers dried quicker than writing inks, and wondered if they could be used in pens. Ordinary nibs choked on the thicker ink, so he replaced them with tiny ball-bearings, hence the generic name 'ballpoint pens'. The thick ink was also less likely to leak.

Sales took off once peace broke out. Competing vendors claimed the 'leak-proof' pens could write for a year, even write underwater, but poor manufacture caused many not to write at all. The ballpoint all but succumbed to consumer dissatisfaction; fountain pens returned to favour. Balls of stronger, more reliable tungsten carbide, introduced in the late 1950s, revived consumer trust. Today's major supplier, French company Bic (founded in 1950 by Baron Bich), sells 20 million 'biros' a day. Fountain pens are now more for status than everyday writing, much like the appeal of early watches (p. 11).

Meanwhile, the typewriter still dominated the office. The best typists made errors; correcting them was messy and laborious. Dallas typist Bette Graham was the saviour, inventing in 1951 what she called Mistake Out, but which later became universally known as Liquid Paper. She mixed up a batch of white tempura paint in her kitchen blender, used it herself to cover up and type over mistakes and sold it to workmates. Her business boomed (after IBM declined to buy her patent) and she sold out to the Gillette Corporation in 1979.

Initially water-based, Liquid Paper dried slowly and did not suit all papers. By the early 1980s, rival products containing fast-drying organic solvents reached the market, the most famous being Wite Out, applied with a brush and drying in seconds. In 1992 Bic bought the rights to Wite Out, soon dominating the correction fluid business as it already did ballpoint pens. It also took over TippEx correcting paper which, when typed over, could erase a typo. The demand for such products was to fall with the advent of word processors, since errors could then be corrected before they ever reached the page.

The Age of Plastics

Wallace Carothers and Many Others

In the modern age of plastics, when man-made materials have become useful, even indispensable, in every aspect of our lives, nylon fills a special place. This iconic product was the first artificial fibre that was totally synthetic. It is made ultimately from minerals such as coal, water and air, unlike the earlier rayon whose starting point was cellulose from plants (p. 152).

First produced in 1938, Fibre 66 (the nylon name came later) was initially used in toothbrush bristles, it was soon seen as an artificial silk, and once war came replaced real silk in parachutes. The global conflict delayed nylon's most famous use, in women's stockings; the words 'stockings' and 'nylons' were later almost synonymous. As nylon is a thermoplastic that softens when heated, clothes made from it can have permanent creases and pleats. Its strength, lightness and toughness see nylon widely used in ropes and many engineering applications.

The Great Polymer Hunt

Nylon flowed from research at the American DuPont company, founded 150 years earlier to make gunpowder, and already the makers of cellophane. DuPont was a leader in the new field of polymer chemistry, itself one of the most important developments on the twentieth century, if measured by the diversity of uses manmade polymers have today. The natural world is full of polymers, long molecules made by joining up many small chemical units ('monomers'), often identical. For example, a molecule of the polymer cellulose, found in cotton or linen, is built up from thousands of units of the simple sugar monomer glucose.

Through the 1920s and 1930s, chemists at DuPont and elsewhere sought to do the same trick with other simple molecules including many made from coal or crude oil. DuPont research chief Wallace Carothers, who led the nylon discovery team, was among those arguing that by carefully choosing the monomers, and even chemically altering them, the resulting polymer could have whatever properties were desired.

Acrylic, Polystyrene, Vinyl

We now find everyday materials based on this polymerisation principle everywhere. Monomers derived from acrylic acid can be joined together in one, two or three dimensions to make acrylic fibres for garments, acrylic paints that dry in minutes rather hours or days (polymerisation here produces a thin sheet rather than a thread), and solid transparent plastics marketed as Perspex, Lucite or simply 'acrylic' (common in unbreakable drinking glasses). Early contact lenses were made from such material (p. 148).

Starting with the monomer styrene, we can produce polystyrene, first manufactured in 1930. Blowing air through liquid styrene produces the polymer as a lightweight foam (styrofoam), used in hot drink cups, packaging and thermally insulating containers. Its tendency to crumble and to blow about makes polystyrene less than ideal environmentally.

Materials known collectively as 'vinyls' are now ubiquitous. The name reflects that they are polymers of molecules of the simple chemical vinyl chloride. First created in 1927, polyvinyl chloride (PVC) is inexpensive, durable, chemically inert, able to be moulded and fire-resistant. No wonder it is so versatile and widely used, in bottles, floor and furniture coverings, shoes and clothing, as well as in plumbing.

Polythene and Teflon

The story of polyethylene (nowadays polythene) shows that not all these discoveries are planned; some just happen. In 1933, two chemists at the ICI (Imperial Chemical Industries) laboratories in the UK were experimenting by subjecting gases such as the hydrocarbon ethylene to very high pressure. On opening the test cylinder they were amazed to find not ethylene gas but a white waxy substance like plastic. This was the first ever polythene and by 1936 ICI could make it in bulk.

Polythene arrived in time for mass production during World War II. There it played a vital role as a lightweight insulator for electrical equipment, such as the newly invented radar sets (p. 205). Nowadays we see polythene everywhere, especially as transparent film in plastic bags and cling wrap, and in durable containers. More polythene is manufactured worldwide each year than any other plastic.

Similar serendipity created Teflon. Starting with the newly created Freon, developed to make refrigeration safer, two DuPont chemists accidentally polymerised the gas by leaving it in a cylinder overnight. First made in 1938, chemically inert Teflon is now famous for its non-stick properties, so much valued in the kitchen and in industry.

The Age of Plastics has its doubters. Polymers are usually resistant to normal processes of decay, and so hang about in the environment for long periods, especially if carelessly discarded. For example, recent decades have seen a backlash against the ubiquitous plastic bag, with consumers encouraged to use paper or cloth bags to carry their groceries. Yet if we consider the impacts from cradle to grave, synthetic materials may prove environmentally sounder than natural ones. The production of paper for bags, for example, consumes very large amounts of water.

The Reactor and the Bomb

Harnessing Nuclear Energy

In terms of political and military consequences, no scientific discovery in history outranks nuclear fission. 'Atomic' or 'nuclear' energy was not invented; the potential had always been there. What we invented were ways to harness atomic energy, releasing it on demand, gradually for peaceful purposes or violently for military ones.

The key discoveries were made through 1938 and 1939, on the very eve of global conflict, by scientists in half a dozen countries, including Germany. The key facts were these. When the cores or 'nuclei' of uranium atoms are bombarded with tiny uncharged particles called neutrons, some of them split in two, releasing energy, a few neutrons and smaller nuclei which were strongly radioactive. This process is nuclear fission. The freshly released neutrons can cause other atoms to split, starting a self-sustaining chain reaction with ongoing release of energy. Compared with ordinary chemical reactions, like those that power dynamite (p. 127), the energy liberated is immense.

There was still some uncertainty. The energy might or might not come out fast enough to be explosive, able to power a bomb. With war looming, that was a major concern. The crucial factor was uncovered by scientists in Britain in 1940. Natural Uranium has two forms or isotopes; the rare U235 and the abundant U238. Neutrons split only U235. To make an atomic bomb, the uranium had to be 'enriched', processed to greatly increase the proportion of U235. That was one possible path to an 'atomic bomb' one that in reality led to the levelling of Hiroshima.

The Plutonium Possibility

There was another way. Years before, Italian physicists, led by Enrico Fermi, sought to make elements heavier than uranium ('trans-uranics') by making uranium absorb neutrons. Such experiments continued in the US after war broke out in Europe, and with the help of many Jewish scientists who had fled persecution, including Fermi himself. Just as fission was being discovered, researchers learnt that one of the newly found transuranics, the ominous-sounding 'plutonium', made by adding neutrons to U238, would also split when hit with neutrons, perhaps even more violently than uranium.

The military implications were clear. If enough plutonium could be created, it too could be made into a weapon. Indeed, the bomb that destroyed Nagasaki was powered by plutonium. There was another benefit. Since the plutonium was manufactured from the 'useless' U238, which made up 99 per cent of natural uranium, the amount of fissile material potentially available rose 100-fold.

Building the Gadget

All that was discovery. Now came the invention, on a scale and with a sense of urgency never seen before. The vast Manhattan Project, charged with creating war-winning weapons from uranium and plutonium, involved top-secret laboratories in a dozen places across the US. It ultimately employed hundreds of thousands of people, including many of the brightest minds of the day, solving an immense array of technical problems on the way to the 'gadget', so-called for the sake of security.

In Chicago, a Fermi-led team built the first nuclear pile on a university squash court. The reactor (our term today) was literally a pile; pieces of uranium distributed among blocks of graphite, the 'moderator' to slow the neutrons down so they would be more easily absorbed by U238, creating plutonium. The build-up of the reaction was controlled by metal rods that absorbed neutrons and were slowly withdrawn. On 2 December 1942, the pile went 'critical'; initiating the first man-made chain reaction. The way was now open to make plutonium, a few milligrams at a time. There could be enough for a bomb (only a few kilograms were needed), given enough time and enough reactors. Had anything gone wrong, a man stood by ready to cut the rope supporting the control roads with an axe. They would have dropped back into the reactor and halted the reaction. Shutting down a reactor is still called 'scramming'. SCRAM is reputed to stand for Safety Control Rod Axe Man.

The other route to an atomic bomb, enriching the uranium mixture in U235, was being pursued elsewhere, with various methods being trialled for this difficult and delicate task. A lot of new physics and chemistry was learned in a hurry, but gathering the fissile material was excruciatingly slow. By July 1945 there was still only enough available for three bombs. One would use uranium, two would use plutonium; one of the latter was needed for a test. With the war in Europe over (the whole Project had been driven mostly by the fear of Hitler having the Bomb), Japan was the target. The Pacific War ended soon after, though the atom bombs did not change the outcome.

In the decades ahead, the fission bomb would give way to the fusion bomb (p. 238) and be married to the rocket to make the fearsome Intercontinental Ballistic Missile (p. 174). Fermi's first reactor would be followed by hundreds more (p. 232).

The Helicopter Takes Off

Igor Sikorsky

The idea for the helicopter (p. 12), supported and driven by large, rotating horizontal blades, was probably inspired by the spinning sails of a windmill. Instead of moving air pushing the sails around, the powered sails would push on the air. Toys operating on this principle had a long history, originating perhaps in China, the rotors set spinning by pulling on a cord. Later models had onboard clockwork motors, even steam engines.

By the nineteenth century 'screw' propellers were pushing ships (p. 99) and were used in early attempts at powered flight (p. 172). But a practical load-carrying helicopter came only in the mid-twentieth century, mostly through Russian-born engineer Igor Sikorsky. Son of a psychiatry professor and a doctor, Sikorsky's early interest in aviation was stimulated by news of the Wright Brothers and Zeppelin airships (p. 165). He became a major innovator, building the first ever multi-engine aircraft, promoting revolutionary ideas like all-metal construction, enclosed cabins with upholstered chairs for passengers, even an onboard toilet. His designs presaged later airliners; his amphibious aircraft made possible airline routes into regions without landing fields.

Sikorsky had migrated to the USA after the Russian Revolution. In 1931 he returned to his early attempts at a helicopter, undertaken in Russia in 1909 aged only 20. The weight of engines kept his creation on the ground. No experimental machine of the time went very far or high, if it flew at all; control was a challenge. The first real success with horizontally rotating blades came with the 'autogyro', devised by the Spaniard Juan de la Cievra in 1923. An ordinary air-screw pulled the plane forward, supported by large unpowered rotors spinning freely.

As with early locomotives (p. 83), progress depended crucially on the power-to-weight ratio of the engines. When more-efficient lightweight engines arrived, along with light, strong construction materials, a number of inventors began to experiment again, including the German Heinrich Foche. Sikorsky had the most success. To stop the whole machine turning in the opposite direction to the rotor, he added a second small rotor on the tail (other designers had used counter-rotating blades). His prototype VS300 first flew in May 1940, with the designer at the controls. A year later, a production VS300 stayed aloft for 90 minutes. The larger R4 made the first-ever helicopter rescue in Burma in 1944. Sikorsky's single-rotor machine remains the dominant design today.

Mastering Microwaves

Edward Bowen, Mark Oliphant, John Randall, Harry Boot, Percy Spencer

The development of radar, particularly in Britain in the 1930s, had a profound influence on the course of World War II (p. 205). Early radar sets used radio waves 10 or 20 metres in length, but those had limitations, especially when the military wanted to install radar sets in fighter aircraft so they could find and destroy bombers attacking British cities at night. Both the size of the gear and the precision with which a radar beam could be directed onto its target depended on wavelength. The new need called for radio waves of about 10-centimetre wavelength (microwaves). The crucial calculations were made by Welshman Edward Bowen, one of the very first to join the British radar effort in 1935.

At the time, microwaves could be produced at a power of only a few watts, nowhere near the kilowatts needed to get reflections from planes and ships kilometres away. The physics laboratory in Birmingham University, headed by Australian Mark Oliphant, soon had the answer. His colleagues John Randall and Harry Boot invented a unique device; it made electrons spiral around magnetic fields in holes bored in a solid block of copper, converting their energy into electromagnetic waves.

Well-connected in defence matters, Oliphant (who also played an important role in the development of the atomic bomb) brought the invention to the attention of military authorities, who seized upon its potential. At its first firing up in early 1940, the 'cavity magnetron' had produced over 400 watts of power at a wavelength of 10 centimetres, and that became 500 kilowatts once engineers at General Electric got to work on it. They turned the temperamental laboratory prototypes into equipment robust and reliable enough to go to war.

Microwaves at War

Microwaves and the magnetron transformed the use of radar. Night-fighters could now easily find their targets in the dark, attacking bombers could navigate and strike their targets using a radar set that generated an image of the ground over which they were flying. The Allied victory against German submarines in the vital 1943 Battle of the Atlantic was largely due to microwave radar. The Americans, for all their expertise, had developed nothing like the cavity magnetron. When alerted to its existence by Bowen and others, so they could undertake further development, American experts called it 'the most valuable cargo ever to cross the Atlantic'. German engineers soon came up with something similar, and technically at least as good, but the equipment was never as well integrated into the overall scheme of defence as the British had done.

Microwaves at Peace

Mastering microwaves, initially a necessity of war, continued its impact over coming decades. Radar soon found all sorts of civilian applications, including air traffic control and weather forecasting. In communications, microwaves, able to be beamed very precisely and to encode large amounts of information, began to replace buried copper cables in carrying large numbers of telephone calls. Such 'microwave relay' chains now cross the continents and send messages to and from orbiting satellites, though on high-volume routes, the even more capacious optic fibres (p. 262) have now begun to take their place.

Microwaves oscillate billions of times a second and have found a vital role in the particle accelerators physicists use to seek the ultimate structure of matter (p. 236). Generated now by high-power machines called klystrons, microwaves are ideal for rhythmically pumping in energy to speed up subatomic particles as they orbit the accelerators, or sprint down a straight track kilometres in length.

Microwaves in the Kitchen

There is of course a more homely use for microwaves. Most kitchens in industrialised nations now have a 'microwave'; indeed it has replaced the conventional ovens in many homes. The heating effect of microwaves was obvious from the start; it arises from molecules, particularly of water, absorbing microwave energy and heating up. Scientists in Birmingham in 1940 could light cigarettes in the beam crackling from the first magnetron. With a focus on, as one wag said, 'cooking the Nazi goose', domestic applications had to await the end of hostilities.

The concept was vigorously taken up by Percy Spencer, an engineer at the American electronics firm Raytheon, who found a chocolate bar in his pocket melted when he walked through a microwave beam. The first production microwave oven, known as the Radar Range, and marketed in the 1950s, used a magnetron similar to those in a radar set, but it was massive and expensive and hardly sold. Both size and price came down through the 1970s, and 'microwaves' took hold, especially when controlled by microchips (p. 267) so becoming very simple to operate. Nineteen out of every 20 households in the USA now have a microwave, with the popping of popcorn one of the most popular uses. The global population of microwave ovens is estimated at 200 million.

Breathing Underwater

Jacques Yves Cousteau, Emile Gagnan

To stay very long underwater you have to be able to breathe. This is easy enough in a submarine (p. 164) or in a diving bell (p. 36), but harder if you want to move freely and independently. The traditional diving suit and heavy metal helmet allowed wearers to walk around on the seabed, but they were still tethered to a ship by a hose through which came life-sustaining air.

By the early 1930s, daring divers were strapping tanks of compressed air attached to a facemask to their backs and heading underwater. But there was a risk. Water pressure increases with depth and threatens to crush the lungs unless the air in them is at the same pressure. So divers had to increase the pressure of the air they were breathing as they went deeper, turning a valve by hand. Unless this control was precise—not easy to ensure if something else was distracting the diver's attention—lungs could collapse and the diver drown.

Among the keen exponents of this hazardous sport of 'free diving', was French naval officer Jacques Yves Cousteau, who had already had some close escapes. He knew that safety could come only through a system that automatically adjusted the pressure of the air being inhaled to the external water pressure. In 1943 Emile Gagnan provided the solution. An expert in regulating valves for gas supplies, he devised a unit with a diaphragm able to balance pressures. When the diver breathed in, water pressure pushed on one side of the diaphragm, opening up the compressed air supply, which then pushed back on the other side of the diaphragm. Air flowed until the two pressures were equal, and then cut off. Breathing out pushed air through an exhaust valve so that the cycle could begin again.

It was Gagnan's vital invention that made possible the aqualung or 'scuba' equipment (self-contained underwater breathing apparatus), and he shared the financial rewards. But in the popular mind, Cousteau got the credit. Indeed, through his use of the new equipment on many pioneering dives, the subject of numerous films and books, he became a household name, opening up awareness of the undersea world to millions of readers and viewers. In the 60 years since, 'scuba diving' has become a popular sport and a great aid to marine biology and underwater archaeology.

Mustard Gas:
Killer and Cancer Cure
The Start of Cancer Chemotherapy

Late in 1917 at Ypres in Belgium the technology of war entered a new phase. For the first time, German forces used a poison gas specifically refined for use as a weapon (earlier gas attacks had used chlorine). Mustard gas, also called Yperite for obvious reasons, was nasty stuff. It was very powerful, needing only low concentrations to be effective, and very persistent, remaining active in the soil or on a soldier's equipment for days or weeks. It had been known since the 1860s, though it does not occur in nature. Its use required the invention of equipment to manufacture and disburse it in large amounts. The other side were not slow to make use of the new weapon; it was a major factor in breaking through the Hindenberg Line in 1918.

Since the gas was absorbed directly through the skin, wearing a gas mask made little difference. Despite a faint, if distinctive, smell, soldiers were commonly unaware they had been exposed until symptoms developed a few hours later. The first organs affected were the eyes, which swelled up and closed over; corneas became ulcerated and died, so victims often went blind. Large blisters appeared on the skin. Mustard gas, actually a liquid dispersed as an aerosol, was more incapacitating than life threatening, since it was believed sick soldiers would more impede an army advance than dead ones. Only 1 per cent of exposure resulted in death, mostly from damaged wind pipes—men choked. Even those who survived were often disabled.

There have been consistent reports of mustard gas being used in conflicts since that time, despite international action to outlaw it, such as by Japan against China during World War II and by the Iraqi government against Kurds in 1988. Despite this grim history, we can note that some unexpected good came from mustard gas. When a stockpile maintained by US forces in Italy was bombed in 1943, a noticeable drop occurred in white blood cell counts in civilians exposed to the gas. Perhaps the gas, or something like it, could treat various forms of lymphoma, a cancer in which white blood cells proliferate uncontrollably. A drug called mustine, similar to mustard gas, was developed post-war, and was the first drug ever used successfully for cancer chemotherapy. Combined with oestrogen it was also used to treat prostate cancer. This extended the use of chemicals against disease first promoted by Paul Erlich (p. 183).

The Father of Hypertext

Vannevar Bush

Open a document on any website, and you will find some words underlined, bolded or in a different colour. Point your mouse at one of those and click, and a 'hot link' opens, taking you instantly into a new document relevant to the keyword clicked. The new document might be a definition of the word itself, the biography of a person named, the full text of a document referred to and so on. That is an example of 'hypertext'—'non-sequential writing; text that branches and allows choice to the reader, best read on an interactive screen'.

Hypertext is not new. In defining a word, a dictionary will often include another word in small capitals, indicating that there is a separate entry for it. Cross-referencing by footnotes adds to the information stock. An early worker on computer-based hypertext was the inventor of the 'mouse', prolific American inventor Douglas Engelbart. In 1968 he introduced his NLS (oN-Line-System), with a large number of documents held in an accessible memory, all cross-referenced and 'hot-linked'.

But the founding visionary of hypertext, and the inspiration for Engelbart and others, was American engineer, scientific statesman and World War II presidential adviser Vannevar Bush. He oversaw most of the technical developments affecting the war effort, including the Manhattan Project (p. 216). In July 1945, around the time of the first atom bomb tests, Bush wrote about the Memex, an imagined machine able to hold an immense amount of information in words and images, but with a mechanism that could link any particular item in the store of information to any other item, as seemed appropriate.

Users moving from one document to another would create a trail of links, able to be retrieved so that they (or others) could follow the same path. Any one document might be reached by a multitude of paths, whereas documents stored alphabetically or numerically or by subject can be reached by only one set of rules. Those rules do not take advantage of the powerful capacity of our minds to make associations, as Bush claimed Memex would do.

Memex was never built, perhaps could not be built, even with the rapid improvement of microfilm, which Bush envisaged as the way to store documents. He described the Memex as the size and shape of a desk, with images projected onto sloping screens for easy reading. Does it sound familiar?

The First Modern Computers?

John Mauchly, J. Presper Eckert

You can get a good argument going over who invented the modern computer. In reality, the features we associate with computers today—electronic, using binary code, programmable—did not emerge all at once. Some early computers had some of these properties but not others. Some used electromechanical technologies rather than electronics; some stuck with 10-digit numbers rather than binary code; some were designed for one given task while others were more versatile. Several nations were involved: the USA, Britain and Germany. But all were invented in the late 1930s or early 1940s, and their first (or intended) uses were military—for cracking codes and undertaking complex calculations (firing tables) to improve the accuracy of artillery fire. But only a few of these machines went through more than one generation.

ENIAC and UNIVAC

Combining these factors suggests that priority as the first modern computer should go to the Electronic Numerical Integrator and Calculator (ENIAC). Construction of this began at the University of Pennsylvania in 1943 under the leadership of John Mauchly and J. Presper Eckert and it went live three years later. This was certainly electronic and could be reprogrammed (though with some difficulty), but it was not binary, continuing to use decimal numbers. ENIAC was colossal in every dimension, weighing 30 tonnes, consuming as much electricity as 50 homes, with 17 000 vacuum tubes and five million hand-soldered joints.

It was ponderous by modern standards, with a 'clock speed' of 5 kilohertz (today's laptops are a million times faster, supercomputers a billion times). Data was fed into the machine by punch cards and came out the same way. Originally intended to figure firing tables, its first calculations supported the development of the hydrogen bomb (p. 238). With upgrades, it was kept running for a decade.

ENIAC was a one-off. Nothing quite like it was built again; the technology was already picking up speed. But it did have offspring in the form of UNIVAC, the first electronic computer to reach the market in the USA (though comparable machines were already on sale in Germany and Britain). The name stood for Universal Automatic Computer, and 'universal' was the right term. From the start, the builders had business needs in mind; customers would be able to assemble an information-processing system to meet their precise needs.

UNIVAC was conceived in 1946 by Eckert and Mauchly, who soon after went into business building computers, but the godfather was the statistics-obsessed US Bureau of Census, which put up the first funding. It is worth remembering the support the US census gave to Herman Hollerith half a century before (p. 153). The money wasn't enough and a project for the Northrop Aircraft Company—a lightweight binary computer to see if test ballistic missiles could be controlled in flight—was signed up to make ends meet. Even so, Mauchly and Eckert were short of cash and early in the 1950s their UNIVAC company passed into the control of Remington Rand, the leading maker of office equipment and adding machines.

UNIVAC at Work

The first machine went to the census office in March 1951 after six years of development. Over time nearly 50 were sold at a cost of $500 000 in 1950s money. The demand came from private-sector firms and government departments preoccupied with numbers: the Prudential Insurance Company, market researcher A.C. Neilson, big companies needing help with their payroll, the military, universities and research laboratories. At its first public demonstration, a UNIVAC successfully predicted the outcome of the 1952 US presidential election. Basing its judgement on previous voting patterns, a UNIVAC hired from the Atomic Energy Commission by CBS predicted victory for Eisenhower over Stevenson with only 1 per cent of the vote counted. No wonder the media were beginning to talk about 'electronic brains'.

Though much smaller than ENIAC, UNIVAC was still a monster. Five thousand vacuum tubes filled a space 5 metres long, 3 metres high and 2 metres deep. But it was twice as fast as ENIAC at adding numbers, (two 12-digit numbers in one ten-thousandth of a second), and had for the first time a significant internal memory, about 12 kilobytes in today's terminology. It was still not binary, working with ordinary decimal numbers. But it was showing the way, particularly by having a stored program—a set of instructions that told the central processing unit (CPU) the steps to take to complete the task. This vital feature was the creation of the brilliant Hungarian–American John von Neumann. The task could be changed just by substituting one program with another (rather than rewiring the machine, as required by the ENIAC).

Already optimists were predicting that the next generation of computers would 'weigh less than a tonne', though few would have conceived a computer that would sit on a desk, let alone on a lap. But that was coming (p. 285).

Cashless Commerce

John Biggins

Today 'plastic money' is almost as common as the real stuff, and often more convenient. Credit cards have transformed the handling of money, impacting heavily on retailing and personal finance. The idea is at least a century old. In the 1890s in Europe, merchants began using cards to administer credit offered to valued customers. The practice gained popularity through the 1920s especially in the USA. Such 'store cards' did not involve the banks directly and could only be used at the issuing organisations, such as merchants, hotel chains or petrol retailers.

Things began to change after World War II. John Biggins of the Flatbush National Bank is usually acknowledged as inventing the bank-issued credit card in 1946. His 'Charge-It' system allowed merchants to bill the bank, the bank then debiting the account of the card-user. The practice spread rapidly through the 1950s. Banks collaborated to develop the widely accepted cards now known as Visa (first issued under another name in 1966) and MasterCard (1967). At the same time some now familiar cards made their debut. In 1950 Diners Club issued 200 customers with a card that could be used to pay bills at 27 New York restaurants, with the customer later repaying Diners. American Express, already profiting from

the issuing of travellers cheques, followed in 1958.

Further advances were sparked by new technology. The introduction of magnetic stripes on cards allowed them to store important information such as identification details. IBM had pioneered this in the 1960s and they were first used to speed ticketing at airports. By the early 1970s, the use of magnetic stripes had been standardised, with resulting improvements in security and efficiency. All cards still have them, though there is increasing use of embedded memory chips (p. 282) to supplement or replace the strip, as they hold much more information.

Rapidly advancing information technology has further boosted the multimillion-dollar credit card industry. Most credit card readers are now online, linked into central computers that can instantly check the customer's status and issue approvals. Fraud has decreased, paperwork has been decimated. Debit cards and EFTPOS (electronic funds transfer at point of sale), available since the 1990s, allow people to use their own money for cashless purchases, with funds moved from their account to the merchant's. We have not yet reached the long-promised 'cashless society', but we have certainly moved a long way down that path.

Sunglasses and Instant Pictures

Edwin Land

The Polaroid 'instant' camera is now obsolete, overtaken by the advance of digital or electronic cameras. Yet for 30 or more years, it offered something no existing camera could—pictures able to be admired (or discarded) within a few minutes of being taken.

American Edwin Land was the innovative mind and driving force behind both the Polaroid camera and the widely used Polaroid filters. The phenomena of light polarisation caught Land's attention while he was a physics student at Harvard (though he left before graduating to go into business). In a light wave, the energy fields vibrate from side to side at different angles as it progresses. Polarised light has all the vibrations lined up.

Early in the nineteenth century, French researchers had found that Iceland Spar, a naturally occurring crystal, filters light according to the angle the vibrating fields take. While still a student, Land put that property into plastic sheeting. Everyday unpolarised light battles to get through a polarising filter, so its intensity is greatly reduced. Land's 'Polaroid' was therefore ideal for sunglasses, reducing glare without affecting colour. Polaroid was also used to separate the two images presented simultaneously in 3D movies so that each eye sees only one image. There were other uses in photography and science. Major firms like Kodak and General Motors bought Land's unique product. It was a goldmine.

For a time, the Polaroid camera promised to be the same. A Polaroid camera does not use Polaroid; the name is simply that of its manufacturer, the Polaroid Corporation. The inspiration for an instant camera was, we are told, Land's three-year-old daughter. While on holiday in Santa Fe, she asked why she could not see the picture of her he had taken a minute earlier. Why not, indeed? Land at once sketched out the many processes needed to develop exposed film automatically inside the camera, so it could eject a finished print within a few minutes.

The first version, making sepia prints, took three years to perfect, reaching the market in 1947. Instant black and white pictures took another two years, instant colour pictures a further decade. Given the complexity of the processes, the technology was remarkable; through the 1970s sales soared. But film for instant movies failed to sell, strong competition coming from the equally new (and equally 'instant') 'camcorders', which saved moving images on magnetic tape (p. 243). Land dismissed the threat, believing that photographic film would always outperform electronics. The ultimate demise of the Polaroid camera settled the argument.

Fathering
the Transistor

John Bardeen, Walter Brattain, William Shockley

The transistor, which has totally transformed electronics in the last 50 years, had three 'parents'. One was the need to replace the large, fragile, wasteful and often unreliable 'valves' or 'tubes' used to amplify electric currents, say on long-distance telephone lines. These looked like light globes and were indeed descended from Thomas Edison's brainchild (p. 137). The second progenitor was the research team, John Bardeen, Walter Brattain and William Shockley, all working at the Bell Telephone Laboratory immediately after World War II.

The third participant in this act of creation was an odd class of materials known as 'semiconductors', with properties between true conductors like metals and graphite, and true insulators like glass or ceramics or dry wood. Semiconductors, typified by elements such as silicon and germanium, let some electricity pass—not a lot, not none, just some. Hence the term 'semiconductors'.

The three researchers were not really a team, even though they later shared the Nobel Prize for Physics. Brattain and Bardeen worked together, but the tough-minded Shockley was mostly a loner. When the first two made a device by pressing three gold contacts onto a piece of silicon (a 'point-contact transistor') and found it could make an electric current stronger, Shockley decided to better them. He went away and made a 'junction transistor', with a tiny piece of one type of silicon between two pieces of another type. This was more robust, easier to make and soon became 'the' transistor. Shockley left Bell and founded a company that was the starting point for Silicon Valley and all that followed.

The secret of the device lay in the odd behaviour of semiconductors. Their mediocre capacity to conduct electricity can be switched on or off by inputting another current, which acts much like a switch . The current through the device can consist of moving electrons (as in a wire) or moving 'holes', places where electrons should be but aren't. By introducing traces of other elements into the silicon ('doping'), we can produce silicon with an excess of electrons (N-type) or an excess of holes (P-type). Shockley's junction transistor had a piece of N-type between two pieces of P-type (or vice versa). A small current in the centre section could control the much larger current flowing between the outer pieces. This effectively amplified the small current just like the old-fashioned triode valve (p. 176).

Capturing the
Third Dimension

Denis Gabor

Our world is three dimensional, with depth as well as height and width. Yet an ordinary photograph loses the third dimension—it does not show depth. We have to infer it from our previous experience.

Hungarian-born British scientist Denis Gabor created a solution to this challenge; the invention won him the 1961 Nobel Prize for Physics. Working in Germany at the time the Nazis came to power, Gabor emigrated, as many did. While working at Imperial College in London in 1947 he invented 'holography' (from Greek meaning 'the whole message'). At the time, Gabor was trying to improve the amount of detail visible on an electron microscope picture. He had previously experienced such serendipity (finding something while you are looking for something else): he had invented a new mercury vapour lamp, while trying to make one using cadmium.

Gabor quickly realised why a photograph has no third dimension. The film records only the intensity of light that strikes it, discarding another vital piece of information, the 'phase'. This can tell us how long the light has taken to reach the film from various parts of the object. The hologram records whether the arriving light wave is at the peak of its regular cycle, at the trough or in between. That is what 'phase' means. This requires the scene be illuminated with 'coherent light', with the peaks and troughs of the light beams initially in step. The only coherent light source then available, Gabor's own mercury vapour lamp, fell short of what was needed. In that sense, holography was ahead of its time. Only once the laser was invented (p. 254) could it flourish, since the laser produces only coherent light.

A hologram does not look like a picture, but its swirling patterns can recreate a three-dimensional image once suitable light is shone through it. Holograms have become useful in many areas, for example, measuring very small distortions of objects caused by stresses. Holograms imprinted on banknotes and credit cards make them much harder to forge. Moving picture holograms are now available, and many artists have explored holography's potential.

Gabor's multi-faceted inventiveness earned him a hundred patents. He was also one of the first 'futurists', arguing that a serious mismatch had developed between technology and human institutions. 'Inventive minds should consider social inventions as their first priority'.

The Music Goes On and On

Peter Goldmark

Recorded sound on cylinders and disks, pioneered by Thomas Edison and Emile Berliner (p. 135), allowed anyone to hear great music, performed by great artists, in their own homes. Radio stations (p. 189) had plenty of material for their programs. Early systems were mechanical, but electrical recording, using electronic amplifiers, arrived in the 1920s. Music sounded more and more like it did in the concert hall or ballroom.

Limitations remained, particularly the length of the recording. The standard 'record' was 10 or 12 inches (25 or 30 centimetres) in diameter, turning 78 times a minute and holding only three or four minutes of music on each side. For that reason popular songs were commonly written to last three minutes. Replaying a complete symphony, let alone an opera, involved a great pile of discs, and much getting up and down to change from one to the next.

A major step forward came in 1948. Working for Columbia records, it took Hungarian-born American Peter Goldmark four years to perfect the 'long-playing record'. On older records, 30 grooves were fitted into each centimetre; the new 'microgroove' technology fitted 100 grooves into that space. The discs also turned half as fast, 33⅓ times a minute, so that each side held six times as much music—20 minutes, enough for two movements of a symphony. The new format enabled a vast expansion in the repertoire of music on record.

Playback was now electrical too; sound quality improved still more. The old steel needle was replaced by a 'stylus', with a synthetic gemstone at its tip to reduce wear and prolong record life. Devotees now spoke of 'high fidelity' music or hi-fi. Unbreakable vinyl (p. 214) replaced the venerable shellac in the discs themselves.

Goldmark and Columbia planned to share their new format with their competitors, but major rival RCA responded a year later with a third format, the '45', smaller and lighter than the old 78 but holding as much music. These could be played on the same equipment as the new 33s, and were especially in vogue for popular music and jazz.

The 78 was now obsolete. Few would have foreseen that 30 years on, the 45s and 33s would themselves become extinct with the arrival of compact discs (CDs), another manifestation of the digital revolution (p. 287).

It's in the Can

Erik Rotheim, Robert Apanalp

The idea of forcing liquid out of a container using gas pressure is an old one. The first liquid so dispensed was soda water (p. 74). In 1790, soon after its invention, Parisians could buy soda water in a device not unlike a modern soda siphon. The soda water was driven by its own carbon dioxide.

In 1927 the first patent on something like the modern spray-can was taken out by Norwegian engineer Erik Rotheim. His invention was commemorated in 1998 by a special stamp issue. Rotheim's metal cans could be filled with a range of liquids and used a variety of propellants, such as methyl chloride. The name 'aerosol can' indicated that the liquid came out in fine mist or 'aerosol'. Over the next 20 years, the cans were put to use as car fire extinguishers (by General Motors) and to spray World War II soldiers in the South Pacific with insecticides to control the spread of malaria.

In a spray-can the propellant is not a gas (which would have to be at dangerously high pressure), but a liquid that boils just below room temperature. Only a little of the propellant evaporates, since the build-up of pressure raises the boiling point. When some of the propellent comes out (along with whatever is being sprayed), the internal pressure drops, evaporation starts again and the can repressurises.

The big year for spray-cans was 1949. American Edward Seymour began marketing paint in spray-cans (it was reputedly his wife's idea). Graffiti artists have been thanking him ever since. Perhaps more importantly, American engineer Robert Apanalp (later a good friend of President Richard Nixon) patented a valve that did not clog. In 1950 his Precision Valve Company made 15 million of them. It is essentially the same valve found on billions of spray-cans today.

Among the early propellants were the man-made CFCs (chloro-fluoro-carbons). These seemed ideal—the right boiling point, non-corrosive, non-toxic, non-flammable. They had been developed for use in refrigerators. However, in 1974 CFCs were first suspected of damaging the ozone layer that protects us from the Sun's ultraviolet rays. By 1985, the ozone layer over Antarctica was thinning and CFCs were blamed. By international agreement (the *Montreal Protocol*), CFCs were phased out over a decade, and mostly replaced with flammable hydrocarbons like butane and propane. Food products such as whipped cream are usually driven out by nitrous oxide.

The Atomic Age Arrives

Hyman Rickover, Walter Zinn

When the first nuclear reactor went 'critical' in Chicago during World War II (p. 216), no one was really interested in the energy it released. The objective was to 'breed' the vital newly discovered element 'plutonium' for use in the first nuclear weapons. The plutonium was created from U238, the more abundant of the two uranium isotopes, not itself a useful source of power.

Nuclear Power Afloat

At war's end, priorities shifted a little. As the Cold War 'arms race' got underway, many reactors were still built purely to produce plutonium. However, others were designed to make use of the heat generated by carefully controlled nuclear reactions, say to boil water, make steam and generate electricity. The nuclear reactor was intended to replace a furnace burning oil or coal. The earliest interest was from the military; US Admiral Hyman Rickover drove the development of compact nuclear power plants for submarines, with the electricity driving motors connected to the propellers. Submarines no longer had to surface to charge their batteries (p. 164), and so became less vulnerable to detection and attack; they could sail long distances without refuelling. The first operational nuclear-powered submarine carried a famous name, the USS *Nautilus*. Launched in 1955, it demonstrated its prowess by sailing under the icecap at the North Pole in 1958.

The world's navies still operate several hundred nuclear submarines. Many were scrapped at the end of the Cold War, during which they played a vital role in the balance of terror, remaining submerged out of sight for weeks or months, ready to launch missiles with nuclear warheads. Only a few types of surface ships, such as aircraft carriers and icebreakers, proved attractive locations for nuclear reactors. These ships require lots of power and need to stay at sea without refuelling for long periods.

Nuclear Electricity

From the early 1950s, engineers began scaling up reactors so that they could make large amounts of electricity for sale to homes, factories and offices. A reactor in the Soviet Union started supplying electricity for 2000 homes in 1954, but the first big (for the day) commercial reactor opened for business at Calder Hall in Cumbria, Britain, in 1956. This would supply electricity to the grid for 47 years until it closed in 2003. Each of the four reactors at Calder Hall could power a town of 40 000 people; many of the reactors to follow would be 10 times as large.

The USA opened its first power reactor at Shippingport in Pennsylvania in 1957. Over following decades many hundreds of such reactors were built in many countries, with France and Japan in particular coming to rely on electricity from reactors because of a shortage of coal and gas. In recent years, cost and safety issues, with public alarm at accidents at Three Mile Island (USA), Chernobyl (Soviet Union) and elsewhere, have eroded both enthusiasm and construction rates. Hopes had been high to start with. Some early proponents had believed that nuclear electricity would prove 'too cheap to meter', a goal always far from reality.

The Plutonium Connection

The link between nuclear electricity and plutonium was always close. The Calder Hall reactors could make useful amounts of plutonium (for weapons); that was their primary purpose for the first decade. That brings us to US physicist Walter Zinn, the first (with his colleagues) to produce any electricity at all from atomic energy. Zinn had been part of the team in Chicago in 1942. Indeed it was he who withdrew the first control rod to set the reactor there running. He was also a key figure in the early design of reactors for nuclear submarines. In 1951 his Experimental Breeder Reactor One (EBR-1) was turned on in the deserts of Idaho, producing enough electricity to light four 200-watt lamps and later the whole building in which it sat. That was not much power of course; indeed the main purpose of EBR-1 was to prove that by converting U238 into plutonium while extracting energy from fissile U235, a breeder reactor could create more fuel than it consumed.

The reactor certainly did that, and for many years there were confident hopes that breeder reactors would solve the world's energy problems. By making plutonium from the non-fissile U235, they would in effect increase the energy available in natural uranium a hundredfold. But high costs and technical delays have seen that promise dwindle away; no breeder reactor today provides commercial electricity, though research continues. There was always another issue: unease at the prospect of a 'plutonium economy', an energy-generating system that depended on producing and exploiting large amounts of plutonium. Plutonium is not only highly toxic, with lethal doses measured in millionths of a gram, but also it can be subverted into nuclear weapons. The risk always seemed too high.

Today's TV:
Colour, Cable, Digital

Television today is a world away from the flickering, monochrome, interference-prone images that people peered at with such wonder in the 1930s, and in greater numbers just post–World War II (p. 206). Many fast-evolving technologies have combined to enhance, even transform, the look of TV pictures and the way they are made and transmitted. These included the use of videotape, microwave relays, satellite communications, transistors, and screens using liquid crystals.

The 1930s had seen the invention of co-axial cable, designed to carry rapidly varying electric currents like those TV required. The first cable networks, some connecting TV stations, others bringing TV to regions with poor broadcast reception, were in place in the late 1940s. Similar cables are one way of connecting to the Internet 'broadband'.

Though many today will not remember or ever believe it, there were dark days in the years BC (Before Colour). TV went to colour first in the USA in 1951, though the transmissions were shut down a few months later when war broke out in Korea. As always, early sets were expensive and penetration slow, but everyone expects to see colour on TV now. Even old black and white movies are often 'colourised'.

Today the momentum is towards High-definition TV (HDTV), though it will be a decade or more before that is well entrenched. The move is little more than a decade old, starting in 1989 with a working party of the International Telecommunications Union, with standard-setting, trials, and equipment development in the 1990s and the first transmissions (by the Home Box Office organisation in the USA) in 1999. An image in HDTV contains two to five times as much information as a regular TV image. This comes from more scanning lines per screen (which sharpens the picture), a wider range of colours (more realism), and a widescreen format (16:9 instead of the old 4:3, which came from the movies).

HDTV uses digital (more accurately, 'binary') data processing, which we now have in music CDs, in place of the old analog (continuously varying currents) process with which TV began. Digital images can be compressed, as in DVDs (p. 293). With redundant information removed, more programs or data streams can be fitted into the same transmission space. This allows refinements such as 'chose your own camera angle' when watching sport. What would Logie Baird and Philo Farnsworth have thought of it all?

Jet Airliners Rule the Skies

From the Comet to the Jumbo

The first aircraft with jet engines, designed by Germany's Hans von Ohain and Britain's Frank Whittle (p. 208) were military fighters that played a small role in the conflict that soon followed. Once World War II ended, designers and builders could turn to the possible use of jet engines in civilian aircraft, for passengers and cargo. New jet fighters and even bombers were also soon in service and the 'sound barrier' broken late in the 1940s.

Jet airliners would be faster and quieter, and with the extra power would carry more passengers further. Costs should fall. The first such endeavour was the British-built de Havilland Comet, which went into commercial service in 1952. It cut the flight time from London to Singapore from two and a half days to one, and heralded the 'jet age'. Celebration was short lived. Over the next two years, three comets suffered catastrophic fuselage failure and crashed with great loss of life, though the tragedy had nothing to do with the engines.

The remaining Comets were withdrawn from service, but other jet airliners remained in development. In 1958 the American Boeing 707 first carried passengers. It was to prove acceptably safe and reliable. Propeller-driven airliners, flying since the 1920s, began to disappear from the skies, though there was later a place for the hybrid 'turboprop' engine in smaller planes, with a jet engine turning a propeller. Aircraft manufacturers concentrated on quieter and more efficient jet engines, fitted to larger and better appointed aircraft, as the cost of international air travel began to fall and demand rose.

Typical of the technology of air travel around the turn of the twenty-first century was the Boeing 747 or 'Jumbo', the most recognisable airliner, and in 2006 still the most plentiful in service, with 1400 operating worldwide. First flown in 1970, 747s can carry more than 400 passengers at 85 per cent of the speed of sound, for distances in excess of 10 000 kilometres without refuelling. All these statistics would have seemed outrageous a few decades earlier. Still larger jet aircraft are being built by Boeing and other manufacturers, as the anniversary approaches of Louis Blériot just making it across the English Channel in 1909. However, no change in technology comparable to the introduction of the jet engine is likely for some decades.

Smashing Up Atoms

Ernest Rutherford, John Cockcroft, Ernest Walton, Ernest Lawrence

Since nuclear physics began at the start of the twentieth century, physicists wanted to know what atoms of matter are made of. Their strategy was to smash open those atoms with minute missiles of various kinds. The first experiments used particles ejected by the radioactive element thorium. Pioneering researchers (including New Zealand's Ernest Rutherford) proved in 1909 that each atom had a tiny core or 'nucleus'; in 1932 they found the uncharged neutron.

Such naturally occurring particles had insufficient power and penetration for more-advanced experiments. The answer was the 'particle accelerator', a machine that applied high electrical voltages to hurl particles (protons at first) down an evacuated runway onto a target. Researchers in the 1930s, starting with Englishman John Cockcroft and Irishman Ernest Walton, used these machines to break many lightweight atoms into bits—to 'split the atom'—and to find unexpected new forms of helium and hydrogen.

Linear particle accelerators were limited in size, and were supplanted by machines more like racetracks than drag-racing strips. Bombarding particles could be made very fast and powerful by being sent thousands of times around a circular path. Machines known as 'cyclotrons', with particles spiralling outward between large highly polished metal plates, were first built in the USA under the leadership of Ernest Lawrence. The immense cost of scaling these up generated more new designs. By the early 1950s, the ring-shaped 'synchrotron' was the preferred machine, and remains so 50 years later.

In the lead here was the huge (for its time) atom-smashing machine grandly called the Cosmotron, which opened for business in the USA in 1952. Its 'racetrack' (an evacuated tube in the shape of a doughnut) was 10 metres across and delivered bombarding particles to targets with more than a billion units of energy. The first linear accelerators had managed only 100 000 units. Higher energies were vital for progress. A more powerful accelerator was like a bigger telescope, seeing finer details within matter. Previously unknown short-lived fragments of matter would be created from the extra energy available.

The pace of innovation continued. Over the next 50 years, accelerators more than a thousand times more powerful than the Cosmotron came on line, costing billions of dollars—more than a single nation could afford. Nuclear physics has become 'big science'. It was at one such laboratory, CERN near Geneva, that the Internet was invented (p. 290).

Cities of Glass

Alastair Pilkington

We commonly associate glass (p. 32) with windows, but a pane big enough to let in significant light was not easy to make. So window glass has traditionally been expensive. The number of windows in a house indicated how rich the owner was (and made a 'window tax' a good money earner for the authorities in times past). Large windows, say for churches, could be made only by joining many small pieces of glass together. Arranging the fragments of glass (some of them coloured) to form a decoration or a religious image made a virtue of necessity.

Techniques to make larger pieces of window glass were devised by craftspeople and glass firms and included pouring molten glass into a mould, letting it set or drawing out a hollow cylinder of glass from the melt (just as glass tubing is made, with air blown in to stop the tube collapsing) then cutting the cylinder down one side and flattening it out with further heat and pressure.

From around 1900 glassmakers could pull out a flat sheet in a continuous ribbon, much as you might draw out a long strand of toffee. This process was faster and cheaper but often left visible ripples and other imperfections in the glass, however carefully it was done. Light passing through it was distorted, acceptable in small domestic windows but not in the larger sheets needed for, say, a shop front. The sheet had to be made smooth by grinding and polishing both sides, a laborious process and wasteful of glass, to produce the much-prized (and expensive) 'plate glass'.

Window glass nowadays is made by a process conceived only in 1952 by Alastair Pilkington of the famous English firm, which had been making glass for 150 years. A decade of development was needed to perfect the process, which involves floating the molten glass on a layer of molten tin. This produced a continuous ribbon of glass of highly uniform thickness, not needing grinding.

Plate glass could now be made relatively cheaply in very large pieces, leading to its growing popularity as a building material and generating the 'glass towers' look of modern cities. New building techniques using cores of steel and concrete to support buildings from within allow the glass walls to be simply hung on the outside, so they do not carry any weight. This means the entire exterior of a building can be made of glass.

Harnessing What Drives the Sun

Mark Oliphant, Edward Teller/Stanley Ulam, Andrei Sakharov

For centuries we wondered what keeps our Sun shining. In airless space it cannot be actually on fire; if it releases heat simply by shrinking under gravity, it would have gone out long ago. We now know that the source of its vast and enduring energy is nuclear fusion. Atoms of hydrogen, the lightest and most abundant element, blend together to form atoms of the second heaviest element, helium. Mass is lost in the transformation, re-emerging as energy—as Einstein allowed for in his famous equation, $E = MC^2$.

Hydrogen comes in several forms: ordinary hydrogen, deuterium and tritium (which is radioactive). While the Sun runs on ordinary hydrogen (far more common than the other isotopes) researchers in their laboratories have had more success in making deuterium fuse. Australian physicist Mark Oliphant, then working at the Cavendish laboratory in Cambridge in the 1930s, worked out the key features of that reaction. He found that deuterium atoms (strictly speaking their cores or 'nuclei') could be fused in two ways, one making tritium, which he discovered, the other making a previously unknown isotope of helium. He tried to extract energy from this reaction by firing one beam of deuterium nuclei into another, but the process took much more energy than it created, so he let it drop. Anyway, nuclear physics at that time was not supposed to be useful.

The 'Super Bomb'

The first success in getting energy on demand from hydrogen fusion was much more violent, the explosion of the first 'hydrogen bomb' by the USA in 1952. In the context of the increasingly chilly Cold War, the hunt was on for increasingly more powerful weapons of mass destruction, much bigger than the atom bombs first detonated near the close of World War II (p. 216). Leading the effort to create such a 'super bomb' was the Hungarian-born US physicist Edward Teller. With his colleague Stanley Ulam, he invented a way to make such a bomb work, a configuration still called the Teller–Ulam design. For a long time, this was a closely guarded secret, though nations like Britain, France, China and Russia were able to work it out for themselves.

Of the thousands of 'thermonuclear' warheads the world clings to today, nearly all use just deuterium, turned from a gas into a manageable solid by chemically reacting it with the metal lithium. Fusion will take place freely only at temperatures of many millions of degrees, such as are found in the core of the Sun. To set the hydrogen bomb going, it is necessary first to explode a fission bomb, using plutonium or uranium, to create the tremendous heat necessary and to compress the deuterium.

These weapons can release unparalleled destructive power in a split second. A World War II atomic bomb released as much energy as ten or twenty thousand tonnes of powerful explosive; a decade later hydrogen bombs liberated hundreds of times more power, equivalent to megatonnes of TNT. Within a few years hydrogen bombs were small enough to be carried by aircraft and missiles, preparing the ground for the 40-year balance-of-terror standoff between the USA, the Soviet Union and their allies.

Fusion Power

While the development of nuclear fusion for military ends proceeded briskly through the 1950s, Oliphant's 1930s vision of controllable energy from the same reaction were followed up much more slowly. There was not the same urgency fuelled by national security and the technical challenge was (and remains) enormous. How could enough deuterium be heated to the necessary millions of degrees and held in some sort of container, able to compress it until the fusion reactions were set going? Two basic approaches have been tried. In 'inertial confinement', pellets of deuterium are bombarded by laser beams of enormous power, so quickly compressed that the fusion would get under way before the pellets blew themselves apart. 'Magnetic confinement' uses very powerful magnetic fields shaped like cages or bottles to confine, squeeze and heat the gas, now electrically charged because of its temperature.

Doughnut-shaped machines following the second path were called 'tokamaks' by the Russians. They set the pace through the 1960s under the leadership of Andrei Sakharov, who was also the godfather of the Soviet H-bomb. Those machines remain the front-runners, as researchers approach the much-sought 'break-even' point, where more energy is generated from fusion within the machine than it takes to heat and control the deuterium gas. With plenty of cheap fuel (though relatively rare, deuterium is abundant in the sea) added to the environmental benefits (fusion does not produce radioactive wastes or greenhouse gases in normal operation), fusion is appealing for large-scale central power plants. But affordable electricity from fusion is still at least several decades away.

The Perfect Dome

Richard Buckminster Fuller

US inventor and philosopher Richard Buckminster ('Bucky') Fuller was widely dismissed as hopelessly utopian. Certainly he built his life around trying to improve the lot of humanity. Admirers likened him to Thomas Edison or even da Vinci, but he did not leave a great legacy of invention, holding only 25 patents.

Fuller had been through some tough times, contemplating suicide in 1927 when unemployed and penniless with a family to support. But he survived to become an advocate of wisely used technology as the path of human progress, with a central role to technologies that were self-sustaining and frugal with resources. He later coined the term 'Spaceship Earth' to stress how we need to husband our resources, as travellers in hostile space must do.

Reflecting on the death of his young daughter in 1922, which Fuller attributed to inadequate housing, convinced him that better accommodation for the poor was a priority. The innovation with which he is most associated, the 'geodesic dome', has been widely used in buildings, though not much for the housing of the poor. Although Fuller patented the dome in 1954, he did not invent it. German Walter Bauersfield of the Zeiss Optica Works had built one in 1922 to serve as a planetarium. But Fuller certainly popularised the design and promoted its advantages.

A geodesic dome is built from many identical elements, usually triangles, which interlink to make an approximate hemisphere or part thereof. Following the line made by the structural elements forms a series of 'great circles', circles whose centres are the centre of the sphere. These are also called 'geodesics', meaning 'globe-dividing', hence the name.

Fuller argued that the geodesic dome is the only structure that gets stronger and cheaper per square metre the larger it is built. Fuller envisaged them made off-site and flown in, though a small one can be assembled from its parts in a few hours. Being essentially spherical, a geodesic dome has the greatest internal volume for a given surface area. The structure distributes stress evenly throughout its members, so it can be made very large. Many of the tens of thousands built since the 1950s as radar domes, sports halls, civic structures and exhibition attractions are 100 metres across. Some span 200 metres with no internal supports. One covers a research station at the South Pole.

Speeding Over Snow and Water

William Hamilton, Joseph Bombardier

The swift-flowing snow-fed rivers of New Zealand are usually shallow, without enough draft for a traditional boat with keel and propeller. As a result, many are not navigable by power craft. New Zealand engineer William Hamilton, born in the high country of New Zealand's South Island, was familiar with its rivers from his youth. His early career focused on earthmoving; he designed new machines for the task, and trained many engineers for war work in World War II.

In the 1950s he returned to the challenge of the shallow rivers, designing a 'jet boat' that ejected water at high speed from the rear of the boat above the water line. His first 4-metre plywood prototype was trialled on the Waitaki River in 1954, though at modest speed. Jet boats are highly manoeuvrable; vanes direct the jet from side to side as required. They can perform 'Hamilton turns', reversing and stopping at high speed within their own length. Though difficult to control at speeds between very fast and dead slow, jet boats are now popular for recreation, flood relief, survey and exploration. The largest are the German-built 120-metre Valour-class frigates used by the South African navy.

Travelling at speed across snow was an equal challenge. Here the most inventive mind belonged to Quebec-born Joseph Bombardier. Though many designs had been tried, including Bombardier's own 'wind-driven sleigh', powered by a Model-T Ford motor car engine and designed when he was only 15, Bombardier quickly saw that tracks like those on military tanks (p. 188) were the way forward. He devised new forms of track, replacing steel with rubber and cotton; stopping one track while the other kept running was a good way to steer. By the 1940s, Bombardier was building 'snowmobiles' with enclosed cabins for up to a dozen passengers. These proved popular with the Canadian police and ambulance services, mining, forestry and exploration companies and served as school buses in winter.

Bombardier galvanised the market in 1958 with the Ski-doo, a one- or two-seater tracked vehicle that resembled an automobile or motorcycle, just as a snowmobile emulated a truck or bus. Instantly popular, these were soon replacing dog teams across Canada and elsewhere, even in Inuit villages, and soon had rivals such as the Polaris. The use of such vehicles remains controversial due to their capacity to damage sensitive parts of the environment that would be inaccessible to most people without them.

The Lure of Solar Power

Auguste Mouchout, Russell Ohr

Nearly all the energy we use comes ultimately from the Sun. Burning coal, oil and natural gas liberates energy trapped long ago in plants by photosynthesis. The Sun's heat evaporates water to fall later as rain and drive hydroelectric systems. Only over the last century or so have we invented ways to capture solar energy more directly.

Sunlight is absorbed by what it falls on, such as the ground beneath our feet, and liberates heat. The quest to harness that heat to do useful work is at least a century old; French academic Auguste Mouchout was one of the pioneers. In the 1860s, concerned that his country was relying on coal, which must surely run out one day, Mouchout built a steam engine that ran on sunlight. Mirrors tracking the movement of the Sun concentrated its energy onto a boiler that turned water into steam. It worked, but not well enough to convince bureaucrats it could replace coal, especially once coal became cheaper. Mouchout went back to his teaching. Today, solar-powered steam engines are a rarity, but rooftop devices on many buildings collect the Sun's energy to heat water to around 60°C for space heating and hot-water supply. Such use is doubling worldwide every five years.

A more versatile approach is generating electricity, which can provide heat, light and mechanical power. A key discovery early in the nineteenth century was the 'photoelectric effect'. When light struck certain newly discovered elements, known as semiconductors, an electric current was generated. The process was not very efficient. Only 1 per cent of the energy of sunlight striking, say, selenium, first used in a 'photocell' in the 1880s, ended up as electrical energy. So the process was mostly a laboratory curiosity, though it was used in light meters to measure how strong light was.

Interest revived in the 1950s, following the invention of the transistor (p. 228) and the LED (p. 256), which also used semiconductors like silicon and germanium. By 1954 Russell Ohr from the Bell Telephone Laboratories had solar cells that converted 4 per cent of sunlight into electricity, and patents to prove it. Better technology has pushed the conversion rate to 25 per cent in the laboratory and 15 per cent in commercially available devices. The initially high costs are coming down, and global production of photovoltaics rose eightfold between 2000 and 2005. Into the twenty-first century, solar cells are increasingly attractive as a source of renewable, non-polluting energy.

Pictures on Tape

From the VTR to the Camcorder

If sound could be recorded for playback on magnetic tape (p. 162), the same could surely be done with moving pictures, since those, too, could be reduced to varying electrical currents. There was certainly a need. Without any means of recording, all early TV programs had to be broadcast live, and were crudely archived by filming the transmission from a screen.

Moving pictures carry much more information than sound does, so the technical challenge was daunting. The early mover was the American firm AMPEX. In 1956 it demonstrated a practical (and saleable) machine that used 5-centimetre tape racing through the recording and playback heads at 40 centimetres a second. AMPEX soon sold 100 of their 'videotape recorders' to enthusiastic TV studios and production houses. New technology provided features like slow-motion and stop-action, but the open-reel VTRs remained too big to sell into homes, despite the efforts of Japanese firm Sony, soon to be a major player.

Video followed the path already pioneered by audio recording. In 1971 Sony introduced the first videocassette, the widely used U-matic. It was followed in 1975 by the (ultimately) ill-fated Betamax. Compact enough for recording and playback at home, the Betamax was still expensive. The next year, Beta had a rival from JVC, the VHS cassette, which was not only cheaper but ran for two hours rather than one.

Battle was now joined in the first 'format war'. The protagonists pushed their technologies, particularly seeking more capacity on each cassette, ultimately reaching nine hours. Movies had to be put out in both formats, adding to costs. Beta probably produced better pictures, but VHS had more industry supporters, and ultimately won out. Beta mostly disappeared from view, though professionals still used it. The dominance of victorious VHS machines, known as VCRs (videocassette recorders) was relatively brief. By 2000, VHS was yielding ground to DVDs (p. 293), and by 2005, most manufacturers had stopped making them.

Meanwhile 'camcorders', lightweight video cameras using small cassettes, were transforming the home-movie market. Cameras using film were soon obsolescent. New formats, recording on tape only 8 millimetres wide, took over, soon joined and surpassed by digital recording on mini-discs. Every advance cut size and cost. Two-package 'luggables' when first introduced, camcorders were becoming truly portable, fitting into the palm of the hand and widely affordable.

The Black Box

David Warren

Whenever a aeroplane crashes, an early priority, second only to retrieving the bodies of victims, is locating the 'black boxes'—the flight data recorder and cockpit voice recorder—with their records of instrument readings and cockpit conversations. This information can be of immense value to investigators seeking a cause for the accident. Nowadays every plane carries them. In 1967 Australia became the first nation to make them compulsory, following an unexplained crash in Queensland in 1960. There is some irony there. Australian David Warren had invented the devices a decade earlier but had found very little interest at the time in making use of them.

Warren worked as a chemist at the Australian Aeronautical Research Laboratories in Melbourne. Following the fatal crashes of several Comet airliners (p. 235) in the 1950s, which were initially unexplained, he explored the possibility of an onboard data and voice recorder, engineered to survive the crash. He reasoned that in many cases the crew would have known what was going wrong, or the instruments would have revealed it. By 1957 he and his colleagues had built a prototype 'flight memory unit' which retained up to four hours of data and voice recorded on steel wire. Trials were successful, but he could not find a manufacturer and there was no support from local aviation safety officials.

Fortunately, his invention came to the attention of an official from the British aviation industry who was much more enthusiastic. Warren was invited to work in the UK with others to improve the device and to house it in a crashproof and fireproof container. The first production model was dubbed the 'Red Egg' from its colour and shape. Sales were soon brisk, and over coming decades many lives would be saved through being able to determine the causes of accidents and take preventative action.

The term 'black box' was always misleading. To aid recovery, the recorders are usually painted red or orange. Requirements are stringent: the instruments must survive very rapid deceleration, crushing, piercing, intense heat and deepwater immersion. Most now are now 'solid state', recording on microchips (p. 282) rather than on tape or wire. Up to 80 different parameters are logged, including speed, aircraft orientation, altitude, engine temperatures and the setting of the controls. Future versions are likely to capture a record from video cameras in the cockpit.

Keeping the Heart on Pace

Wilson Greatbatch

We rely on our hearts—if they stop for long enough, we die. Yet the heart is not complicated, being essentially a muscle that serves as a pump, sending a pulse of blood into our arteries every second or so. Some heart problems can be traced to a faulty heartbeat. Nervous impulses, generated in a piece of heart tissue called the 'pacemaker', fail to drive the heart in a reliable rhythm. The solution is the artificial pacemaker, able to send regular pulses of electric current into the natural one if it is failing.

The idea of using electricity to shock the heart muscle back into regular action is an old one; nineteenth-century physicians tried to induce the needed currents from outside the body. The first to try this internally was probably Canadian engineer John Hopps, around 1941. Cables were run through the chest wall into the heart tissue, but the pulse had to be generated from outside from equipment the size of a modern TV, and the shocks delivered were painful.

The implantable artificial pacemaker, carried within the bodies of millions of people with troublesome hearts, was invented by American bio-engineer Wilson Greatbatch around 1958. There

was some serendipity. While building a device to record heart sounds, he stumbled upon a circuit that produced regular pulses like heartbeats. It was small enough to fit inside the patient's chest. Having generated interest among heart surgeons, Greatbatch begin handcrafting pacemakers in his barn using his own funds, later quitting his job to concentrate on the device. Once the pacemaker had been implanted, the user was soon unaware of its operation. Around 1970 Greatbatch matched his new device with a small lightweight lithium battery to power it, similar to the batteries now widely used in mobile phones and laptop computers. With circuits driven by transistors, the pacemaker drew so little current that batteries lasted for years.

Modern pacemakers have the added advantage of microprocessor (essentially a miniaturised computer) control. They are able to respond more flexibly to changes in the heartbeat, including detecting and dealing with fibrillation or tachycardia, when the heart beats very fast and inefficiently. In 1983 the American National Society of Professional Engineers rated the pacemaker as one of the ten most important engineering contributions to human society in the previous 50 years.

Hard Drives for Computers

Reynold Johnson

An early impediment to the growth of the computer industry in the 1950s (p. 224) was getting rapid access to the data needed for the calculations. Early machines used punched cards, pioneered half a century earlier (p. 153); newer computers were storing data on magnetic tape, like that used to record music (p. 162). Both methods were slow. Cards took time to prepare and to run for the reading machine; tape had to be spooled through from beginning to end. Neither could give quick access to any particular piece of information. As a result, calculations were slower and costlier than they should have been.

IBM, the big player in large computers, set one of its leading engineers, 20-year employee Reynold Johnson, a task: devise a memory that could hold as much data as 50 000 punched cards and access any particular piece in a second or less. With his hand-picked team, and in a new laboratory near what would become known as Silicon Valley, Johnson came up with a solution in two years. It was simple, but high risk: discs of magnetic material, with data recorded and accessed by sensors floating less than a millimetre above the rapidly spinning surface and in danger of crashing into it.

It was also massive: it weighed a tonne and occupied several cubic metres of space, with 50 large aluminium discs spinning 20 times a second. By modern standards, it did not hold much data, only 5 megabytes, much less than a single CD, but that first 'hard drive', launched in 1958, made the modern multi-billion-dollar computer industry possible. As with every area of computer hardware, progress since has been measured mostly by miniaturisation. A typical desktop computer nowadays has a hard drive the size of a paperback book, holding 20 000 times as much data as Johnson's monster, but operating on essentially the same principles.

Johnson had a prolific mind, but a somewhat retiring disposition. His portfolio of 90 patents included the video cassette, devised while he was on loan to Sony (p. 243), and a way of automatically reading cards that have been simply marked with a pencil rather than punched with a hole (like those used in multiple-choice exams). The latter was based on an idea Johnson first had as a high-school science teacher. IBM was so impressed they bought the system and offered him a job. It was a good move.

Scanning the Womb with Sound

Ian Donald

'Ultrasound' refers to soundwaves of 20 000 cycles per second or more, too high for the human ear to detect. Ultrasound waves have applications in nature and in industry. Bats use them to navigate in the dark. Generated by crystals of quartz or other minerals set vibrating rapidly by electric currents, ultrasound is used to find hidden defects in metal structures. It is also the basis of 'sonar', which can detect underwater objects such as submarines or reefs, by the sound-waves reflected.

In medicine, ultrasound was first applied as therapy, using the capacity of the waves to heat tissues or to break up objects like gallstones. A few researchers tried to peer inside the body by measuring how soundwaves passed through structures like the skull. They slowly developed a capacity to make crude images by collecting the soundwaves reflected, bent or absorbed by soft tissues, including tumours.

Ultrasound has had a profound influence in obstetrics and gynaecology, particularly in imaging and monitoring the child in utero. Here the key figure was Ian Donald, of Scottish ancestry and in the 1950s newly appointed to the chair of midwifery in Glasgow. His inspiration came from several directions: war service had taught him something about radar and sonar, and he had a longstanding interest in gadgets and machines. A lecture in London given by one of the other ultrasound pioneers quickly alerted him to what ultrasound scanning could do in obstetrics and gynaecology. His first investigations imaged anatomical specimens such as fibroids, cysts and tumours with an ultrasound metal flaw detector at a shipyard, following a visit made possible by the grateful husband of a patient.

Early trials on patients were disappointing, the work greeted by colleagues with a mixture of scepticism and ridicule. He turned the corner when his primitive scanner found a large operable ovarian cyst in a woman previously diagnosed with inoperable cancer. Attitudes quickly changed. In 1958 Donald was able to announce his findings in *The Lancet* under the title 'Investigations of Abdominal Masses using Pulsed Ultrasound'. From then, as he later reported, there was no turning back. Ultrasound in obstetrics, gynaecology and elsewhere in medicine soon became mainstream.

Donald was strongly opposed to abortion other than on strictly medical grounds. Showing a mother images of her unborn child was one way he sought to discourage termination of the pregnancy.

Making it Stick

Harry Coover, Spencer Silver

Making something stick to something else was a challenge for early craftspeople, and after them for engineers and manufacturers, for millennia. The many solutions included natural gums and resins, boiled starch ('flour and water paste'), fish glue (patented in 1750) and adhesives made from casein, found in milk. For tougher jobs, adhesives dissolved in acetone (nail polish remover) were popular, such as the long-familiar Tarzan's Grip. Such 'drying adhesives' set *as* the solvent evaporated.

In the 1940s Harry Coover of the photographic firm Eastman Kodak was busy with war work. He wanted to use threads of a chemical called cyanoacrylate as a sort of artificial spider silk to make cross hairs for a gunsight. It was a disaster: the cyanoacrylate stuck to everything. Years later he realised what he had. In 1958 Eastman Kodak went to the market with Eastman's 910 Adhesive (not an appealing name), later called Flash Glue because it stuck so quickly (within a minute, even less with some additive) and then Superglue because it stuck so strongly. A 2-centimetre-square patch could hold up a tonne. Soon every home had a tube and the demand from industry was high.

Superglue (aka Crazy Glue and other names) works because it contains small chemical units (monomers) that join together when a little water is present (as it always is) to form long, strong polymer chains. These intertwine to make an all-but-unbreakable adhesive mat. Advanced forms of the glue are now approved (since 1998) for medical use, such as closing wounds and surgical incisions rather than using stitches.

At the other end of the scale, Spencer Silver, employed by Minnesota Mining and Manufacturing (3M), was trying to make a strong glue but ended up with a weak ('low-tack') one, barely enough to hold two pieces of paper together. But it did not dry; however long you left it, it still stuck, sort of. He wondered what use there might be for it, if any. His colleague Arthur Fry found one, to put bookmarks in his hymnal at church choir practice. Ungummed slips of paper fell out; those with the usual glues could not be removed. The new glue was ideal. Today we call slips of paper with a strip of Silver's glue on one edge Post-it Notes. They hit the market in 1980. Where would we be without them?

The Marvellous Microchip

G.W. Drummer, Jack Kilby, Robert Noyce

Information technology began with the invention of the transistor (p. 228), but that was just the first step. British engineer G.W. Drummer clearly saw what was needed next. Electronic circuits remained a tangle of components—resistors, capacitors, switches, transistors and diodes, all wired together. The circuits might not perform simply because some of the joints were incorrectly soldered. In 1952 Drummer proposed something much better:

> With the advent of the transistor and work on semiconductors generally, it now seems possible to envisage electronic equipment in a solid block with no connecting wires. The block may contain layers of insulating, conducting, rectifying and amplifying materials, the electrical functions being connected directly by cutting out areas of the various layers.

Drummer was on the right track, but the technology of the day was inadequate; possible sponsors were hard to convince. In 1957 his first crude model of a 'solid circuit' (the modern term 'integrated circuit' or IC came later) did not actually work.

Across the Atlantic, the military, in particular, was keenly interested in wider use of transistors and similar devices.

Modern weaponry used a lot of electronics. A B-29 bomber carried a thousand glowing vacuum tubes in its radios, radars, navigation systems and control devices. Transistors were rugged, light, compact and cool; this promised smaller weapons that were easier to transport and more reliable. Miniaturisation was the path forward and the integrated circuit led the way.

Jack Kilby, an engineer at Dallas-based Texas Instruments (TI), patented the first real solid circuit in February 1959. A month later TI announced to the world the 'development of a semiconductor solid circuit no bigger than a matchbox'. Even the everyday media took notice. An aggressive pioneering firm, TI would always have been hard to beat to this particular prize. Since 1954 it had been manufacturing transistors from heat-resistant silicon rather than germanium. The low cost of the new transistors made possible the first transistor radio.

Kilby's 'black boxes' were solid enough. A number of wires emerged from a metal case to receive incoming currents and deliver the results. He had found ingenious ways to solve problems that Drummer had foreseen, making capacitors or resistors out of semiconductor material and isolating the various components so they did not interfere with each other. All this would one day win him a Nobel Prize.

Yet the integrated circuits were laboriously handcrafted; components were soldered in place using microscopes and tweezers. The results were costly and unreliable. Furthermore, how far could the process go? Building a circuit with 100 parts was not difficult using Kilby's method, but could it be done with 1000 components or 10 000?

The breakthrough came in Mountain View, California, 80 kilometres south of San Francisco, close to Stanford University. There, physicist Robert Noyce ran a company called Fairchild Semiconductor. William Shockley, co-inventor of the transistor a decade earlier, ran a similar business nearby. These were among the first high-technology residents of the Santa Clara Valley, famous for its fruit trees, later to become universally known as Silicon Valley.

Noyce and the equally brilliant Jean Hoerni had the answer to the IC challenge by early 1959. It began with a new way to insulate the various elements in the transistor using a layer of silicon dioxide, found in nature as silica, the major component of ordinary sand. Hoerni had already invented the 'planar' process, producing flat transistors with no protruding parts. Noyce went a step further. Several transistors could be embedded in a slab of silicon, so it was easy to transform the material between

them into the resistors and capacitors needed to complete the circuit, and then to add narrow strips of metal to interconnect the parts.

The whole thing was built up layer by layer by engraving patterns on the silicon base and filling the cavities with deposits of N-type or P-type silicon, insulators or interconnectors as needed. The process was essentially photographic. Masks with variously shaped holes defined areas to be cut away or built up. Chemicals that hardened when struck by light protected parts of the chip surface to be preserved, while surrounding regions were etched away.

Innovation in microchips has continued without slackening to the present day. By shrinking the size of the photographic masks (and working with light of shorter wavelength) chip makers have packed ever more transistors and other components onto a chip. Spaces that once held a few transistors now hold many millions. Progress has been charted by Moore's Law, a rule of thumb that the number of components that can be fitted on a fingernail-sized piece of silicon doubles in less than two years (p. 282). The resulting thousand-fold increase every decade underpins the explosive growth we have seen in the power of microchips to process and store information, and all the transformations that has brought.

Riding on a Cushion of Air

Christopher Cockerill

Resistance encountered by ships ploughing through the water chews up energy and restricts speed. Innovative minds, the eighteenth-century philosopher Emmanuel Swedenburg among them, have visualised a boat moving just above the water, reducing friction and saving energy. But attempts to build such craft lacked a suitable source of power that would keep them aloft. Human effort was not enough.

The modern version of the idea, the hovercraft, was due to British engineer Christopher Cockerill (later Sir), who had spent the war years working on radar (p. 205). He also ran a boatyard in the Norfolk Broads. If a craft could be supported by an air cushion above water, he thought, the same could happen over land. Ships could simply come ashore to disembark their passengers and cargo; land vehicles could pass easily over uneven and boggy ground. His first experiments on such a 'ground effect' vehicle were primitive, using a few food tins, an industrial air-blower and scales to measure the lift. But he proved the principle.

Encouraged by Cockerill, who saw military potential in the machine, the government contracted the Sauders–Roe company on the Isle of Wight to build a full-sized prototype. This first went to sea in 1959, the event reported to an amazed British public. It later crossed the English Channel in two hours, 50 years to the day after Louis Blériot (p. 173). Early performance was not impressive, limited to almost flat seas and blocked by even moderate obstacles. The addition of a flexible rubber skirt around the craft to contain the air cushion improved obstacle and wave clearance tenfold. The first machine to demonstrate this technology, the SR-N1, is now in the London Science Museum.

Despite its ingenuity and the publicity gained, the hovercraft has not been a great commercial success, at least not at the big end of the market, though some have found military use. Fuel consumption being high, appeal has been limited by rising oil prices. Few large passenger-carrying hovercrafts had been built, and major passenger routes, such as across the Channel from the late 1960s, have since closed. Hovercrafts have been replaced by newer technologies to reduce water resistance—wave-piercing catamarans and hydrofoils that lift most of the hull above the water on struts operating like aircraft wings. Nowadays, hovercrafts tend to be used for specialised purposes, such as delivering mail, freight and passengers along frozen rivers in Alaska. Interest remains high among hobbyists.

The Promise of the Fuel Cell

William Groves, Francis Bacon

Most people remember how close the Apollo 13 moonshot came to disaster. The spacecraft relied on fuel cells for its power, combining hydrogen and oxygen to make electricity (and drinking water). An explosion in one of the oxygen tanks almost ended the mission.

Their use in late twentieth century spacecraft might suggest that fuel cells—where two chemicals react to directly produce electricity, much as in a battery— are a new idea, but the concept is as old as electricity. From 1800 we knew that passing an electric current through water breaks it up into hydrogen and oxygen. How then to make hydrogen and oxygen combine chemically and generate electricity? Around 1840, Englishman William Groves tried it with sealed gas bottles and platinum electrodes. Current flowed, water formed, but very little and very slowly. Fifty years later, two German chemists tried combining air and coal gas (same idea, different chemicals). Success was again limited, though they did invent the term 'fuel cell'. The combination of the new internal combustion engine, which could burn coal gas, and the realisation of how difficult the problem was put further research on the backburner.

Interest revived in the 1930s. British chemist Francis Bacon, descended from his seventeenth-century namesake (p. 25), used cheaper materials, say, nickel instead of platinum. He certainly made current flow, but he took until 1959 to make a hydrogen–oxygen cell produce useful power, say, 5 kilowatts to run a welding machine. The cells needed gases at very high pressure and quite high temperature, and efficiency remained low; much of the gas did not combine.

Other inventors and manufacturers were now tinkering, but it took a decision by the American space agency NASA to use Bacon-style fuel-cell technology in its spacecraft to drive the technology forward. Some predicted a rapid uptake of fuel cells (p. 260), but progress remains relatively slow even today, despite the fuel cell's advantages. It can use any sort of fuel and does not waste energy. At best, a coal-fired power station converts 50 per cent of the energy in the burning coal into electricity. In a fuel cell, it should be almost 100 per cent.

Work continues on many different designs, now encouraged by rising oil prices. Some schools and hospitals have large stationary fuel cells, and major automobile manufacturers have plans. As technical problems are solved, and costs come down, fuel cells will be big players in the drive for clean, affordable power.

Superfibres

Joseph Shives, Stephanie Kwolek, Wilbert Gore

The 'golden age' for inventing 'synthetics'—materials that do not exist in nature—was the 1930s, when we were first introduced to nylon, polythene and synthetic rubber (p. 214). But the hunt goes on for new forms of 'polymer'. The chemical name of a synthetic material includes the word 'poly-' somewhere. Huge numbers of tiny, identical chemical units ('monomers') are linked together into fibre and sheets. By choosing the right monomers, chemists can give the resulting polymer almost any properties, as the following examples show.

Spandex

Also known as Lycra or elastane, spandex was invented in 1959 by Joseph Shives at the US chemical firm DuPont, home of nylon and many others. This remarkable fabric combines the durability of polyurethane (as in floor varnish) with the flexibility of polyethylene (as in clingwrap). It can stretch to five times its normal length without breaking, always coming back to its original shape. Lightweight, soft and supple, spandex is popular in sports and exercise gear, swimsuits, wetsuits, skipants and the like. Comic-book superheroes (and especially superheroines) usually wear spandex when they appear on screen.

Kevlar

Also from the DuPont stable is Kevlar, which makes fibres and fabrics of extraordinary strength. Kilogram for kilogram, Kevlar has five times the tensile strength of steel. Hence its most famous use in bulletproof vests and helmets, but it adds reinforcement in many other stressful situations, such as in car tyres. Its resistance to heat makes it valuable in fireproof garments, where once asbestos was the first choice. Invented in 1961 by Stephanie Kwolek, Kevlar is an 'aramid' fibre (short for aromatic polyamine) and therefore a cousin to nylon. Kevlar fibres are spun much the way a spider spins its web.

Gortex

Gortex is named after its inventor Wilbert Gore who did *not* work for DuPont. It has been used since the 1970s in outdoor clothing and tents, as it is windproof and waterproof, but also 'breathes'. Unlike a plastic raincoat, it allows water vapour to escape, so uncomfortable humidity does not build up inside the garment. The secret is a thin layer of a polymer related to the non-stick Teflon, and so to the CFCs once used in refrigerators and spray-cans (p. 231). This has more than a billion tiny pores in each square centimetre, each much too small to let a water drop pass, but bigger than water vapour molecules.

The Amazing Laser

Thomas Maiman

Lasers are everywhere in modern technology. Yet initially they were just a laboratory curiosity, 'a solution in search of a problem'. The first laser (an acronym for Light Amplification by the Stimulated Emission of Radiation) was built in 1960 by American Thomas Maiman, mostly to prove that it could be done. He used a rod of ruby, which sounds expensive and exotic, but lasers today use a wide range of solids, liquids and gases as the 'medium' to generate their unique radiation.

A laser's talent lies in making the atoms or molecules in the medium behave in exactly the same way and at the same moment. When stimulated, all the particles release energy (in the form of light) simultaneously. That makes the light coherent, with the peaks and troughs of the light waves all lined up. And they all release precisely the same amount of energy, so that all the light is the one pure colour. (The radiation can be invisible infrared or ultraviolet, rather than visible light, but the same properties apply.)

Coherent light has very little tendency to spread out, unlike the beam of a torch, and it can be focused into a very tight spot, less than a thousandth of a millimetre across. That makes lasers ideal for detecting the tiny marks that code information in the grooves of a CD or DVD (p. 287/p. 293). Tight focussing lets a laser deliver its energy into a tiny area, able, for example, to burn very tiny and perfectly formed holes in metal for precision engineering. Coherent light is also a vital need in holography, a technology to give images a 3-D look (p. 229).

The precise colour of a laser beam means that many can be sent together down an optic fibre without confusion, greatly increasing the fibre's capacity to carry data (p. 262). It also let us measure distances with great accuracy, by timing how long a pulse of laser light takes to travel there and back. By bouncing laser beams off orbiting satellites, we know for sure that the continents are moving relative to each other at a few millimetres a year. Most likely, your builder or carpenter now uses a laser device to measure up the job.

All these are powerful applications. But in 1960, Maiman and his colleagues were not thinking about them. They just wanted the laser to work.

Revolution in Reproduction

Russell Marker, Carl Djerassi, Gregory Pincus

Birth control is an ancient human need, for example limiting population growth in hard times. Solutions have included abstinence and interrupted intercourse, intra-uterine devices, 'tube-tying', condoms—greatly improved following the discovery of vulcanised rubber (p. 89)—and abortion. At one time, women in Europe soaked woollen tampons in vinegar or lemon juice to increase the acidity of the vagina and kill sperm.

Understanding the chemistry of female reproduction, beginning in the late 1930s with studies of rabbits, allowed a more scientific approach. We soon knew that doses of hormones like progesterone at the right time could block release of eggs by the ovaries, preventing conception. The first challenge was getting enough of the hormone for widespread use. In 1947 American chemist Russell Marker began extracting large amounts of progesterone from a wild Mexican yam, but there was another immediate problem. The natural hormone could not be swallowed, since digestion destroyed it. Its human use required several painful injections daily.

Forms of progesterone that could be taken by mouth needed further chemical manipulation. Carl Djerassi at Searle, among others, came up with variants like norethindrone and norethynodrel. These would find use in early forms of the contraceptive pill, though at the time their discoverers had other uses in mind. American doctor Gregory Pincus made the connection to contraception in the early 1950s, encouraged by birth-control activist Margaret Sanger (who had coined the term 'birth control'). Pincus first reported success in 1955. Following large-scale trials in Puerto Rico, the first effective birth-control medication went on sale in the USA in 1960. Early high-dose pills raised health concerns, particularly for their effect on blood vessels; lower doses of the hormones were soon found to be both safer and equally effective.

Few, chemical inventions have had the social impact of 'the pill'. Some commentators place it alongside fire and the wheel in importance in human history. By disconnecting sex and reproduction, it overturned long-standing mores. By giving women more control over reproduction, and reducing the burden of unwanted children, it meshed with women's rights and feminist movements in the 1960s and beyond. Relationships between the sexes were changed forever. But there have been downsides. Sexually transmitted diseases increased, as more people, especially younger ones, now had multiple partners and used condoms less often, relying on the pill to prevent pregnancy.

LEDs to Light the Way

Nick Holonyak

In our quest to make the darkness light, we have tried all sorts of technologies, from campfires to candles to oil lamps, from electric arcs to incandescent globes (p. 137) to neon lights to fluorescent tubes (p. 170). The ideal light source would be cheap, efficient, robust, long lasting and versatile. Light-emitting diodes (LEDs) may provide that ideal.

In the history of LEDs, the first name we encounter is American Nick Holonyak. He produced his first LED, which made red light, in 1962 while working for General Electric. An LED is a cousin of the transistor (p. 228). Semiconductors are impregnated ('doped') with impurities to create an excess of negatively charged electrons (in N-type semiconductors) or positively charged 'holes' (in P-type). As a current passes across the junction between N-type and P-type material, holes and electrons combine to liberate energy as light. The colour of the light depends on the material used, and many compounds have been tried. The easiest radiations to produce, and the first to reach the market, were red and infrared (heat). Yellow, green, blue and even ultraviolet light came later (and are more expensive). White light comes from mixing various colours or by making suitable materials glow (fluoresce) by bathing them in ultraviolet light from an LED.

LEDs seem to have all the advantages. Individual units are small but intensely bright and can be clustered for increased light output. LEDs are long-lived; a typical LED will last 10 years, far longer than other light sources. LEDs give off much less heat than incandescent globes (that is, they are twice as good as turning electricity into light) and are more compact and robust than fluorescent tubes (which are, however, more efficient). The impediment remains cost, and that is coming down rapidly as manufacturing techniques improve.

You are already encountering LEDs every day—in your TV remote control (which uses infrared) and your optical mouse, in the dashboard of your car, in the increasingly common, lightweight message display (the one in Times Square in New York is 36 metres high, in new-style traffic lights and flashlights. In time they are likely to replace all other forms of lighting for everyday purposes. Car headlamps are likely to be the next application. And new generations of technology—light-emitting transistors and transistor lasers—are still to make their mark.

Communicating by Satellite

Arthur C. Clarke

The first communication satellites ('comsats') were not lifted into orbit until the 1960s, but the idea was at least 20 years older.

In a 1945 issue of *Wireless World*, science-fiction author Arthur C. Clarke proposed a system of three satellites, equipped with radio receivers and transmitters and each in 'geosynchronous' orbit. Circling the earth at an altitude of 36 000 kilometres above the equator, such satellites would pass once around the Earth from west to east each 24 hours, keeping step with the rotation of the planet beneath and appearing from a viewpoint on the ground below to hang stationary in the sky.

Clarke's vision needed three satellites so that each was simultaneously in view of the other two, as well as in sight of ground stations. The ground stations would transmit messages upwards for relay around the system and then for transmission downwards again to destinations way out of sight of the point of origin of the message. Clarke drew his inspiration from the experiments with rockets carried out in Germany and the USA in the 1920s and 1930s, culminating with the use of such machines as weapons of mass destruction in the closing months of World War II (p. 175). A decade after he wrote, the Soviet Union placed the first artificial satellite into orbit in October 1957 (*Sputnik I*), opening the way for space communications. The start was modest. In 1958 the US Army's *Score* satellite had only one voice channel but could transmit messages directly or store them for a later playback.

Clarke's idea certainly had appeal. A network of comsats would be an advance on the practice of sending messages across land areas by microwave relay (p. 219). Tall towers with transmitting and receiving antennas are built in sight of one another. Clearly we cannot place such towers in the middle of the ocean. A satellite is therefore like a microwave tower tall enough to see both sides of, say, the Pacific Ocean, and relay messages between stations on either side. Furthermore, a satellite can send a message onward to any receiver within its field of view, not merely to a single station as in a conventional microwave relay. The cost of international phone calls would fall sharply as capacity grew.

In 1960 the USA launched *Echo*, an aluminium-coated balloon 30 metres in diameter, able to reflect radio signals beamed up to it from Earth. Soon after, experiments were underway with 'active' comsats such as *Telstar* and *Relay*, equipped with low-noise radio receivers and transmitters to amplify the oncoming signals and send them onward. These experimental rigs could carry hundreds of voice calls simultaneously, and that capacity could be blended to deliver a complete television signal, video and audio. But they were in low orbit, only a few hundred kilometres above the ground and therefore appeared to move swiftly across the sky. Ground stations had to track them mechanically in order to remain in contact, and they did not stay in sight for long.

Much closer to Arthur Clarke's vision was *Syncom II*, the first geosynchronous communication satellite, launched by the USA in 1963 in time to relay live coverage hour by hour of the 1964 Olympic Games in Tokyo to viewers in the USA. For the first time, the phrase 'live via satellite' appeared on TV screens. Nowadays, of course, we are so used to seeing things as they happen on the far side of the world, from wars to royal weddings to tennis matches, that we no longer remark upon it (and the banner has long disappeared). A global network of satellites in contact with thousands of ground stations in many nations sustains an intimate and immediate worldwide exchange of experience through words and pictures that led one early commentator to suggest we were already in a 'global village'. Consider the reach and impact of a global TV news network like CNN, only possible in the satellite age.

The power of a network of comsats to draw the global community together was first demonstrated in 1967 with the program 'Our World'. Short segments showing facets of everyday life in countries around the planet were carried simultaneously by satellite and relay to hundreds of millions of viewers, stitched into a two-hour celebration of diversity. We saw babies newly born on four continents. In a London recording studio, the Beatles sang 'All You Need Is Love', a few minutes later the first tram of the day rattled out of a shed in Melbourne. The response was much like that of the first viewers of the first moving picture show staged by the Lumière Brothers (p. 158). The subject matter was mostly familiar; the means that delivered it to our eyes and ears was extraordinary and unprecedented.

How Fast Can a Train Go?

The Shinkansen and the TGV

The first successful steam locomotive, George Stephenson's Rocket (p. 92), could manage 20 kilometres an hour. That outpaced any other mode of travel at the time over a long haul, and was quite fast enough, thank you, for most passengers. Over the next 150 years, steady if incremental improvements in engines, tracks and railway operations, combined with increased willingness by the travelling public to move at a faster pace, saw speeds reach five or eight times that of the Rocket.

Only France and Japan were thinking seriously about new designs and systems that would go faster. Japan's Shinkansen ('new trunk line'), popularly called the Bullet Train because of its shape, began running between Tokyo and Osaka in 1964, in time for the Tokyo Olympics. Seventeen years later, but still well ahead of anyone else, the first French TGV ('train à grand vitesse' or 'very fast train') began plying between Paris and Lyon.

Not everything about these trains was new. The newly built rolling stock was streamlined to cut wind resistance, but was still powered by electricity from overhead wires, and still carried by steel wheels on steel rails over conventional (if often newly laid) track-bed. The French and Japanese were simply pushing the existing technology as far as it would go.

That was quite a way. The speeds were impressive, averaging over 250 kilometres per hour over a journey of several hours, and reaching over 500 kilometres per hour in trials. Travel was comfortable, departures punctual, environmental consequences minimal and the safety record impressive, especially once the tracks were fenced off. Similar train lines began to appear in other countries, and the lines could compete with air travel over short distances, being cheaper, taking less time door to door and consuming less energy.

Was that enough? Some thought not and were already experimenting with the next-generation technology, a major leap toward 'magnetic levitation' or 'mag-lev'. No longer would trains run on rails. They would float above them, held up, guided and even driven forward by the interaction of electromagnetic fields between the track and the train. With no motors, the trains would be much lighter and more economical to run. No mag-lev train is yet in service, but the Japanese, active since 1970, have reached nearly 600 kilometres per hour in trials. Even higher speeds, approaching 1000 kilometres per hour, would be possible with the train running through an evacuated tube.

1964

Looking Ahead 20 Years

How Good is Technology Forecasting?

While we might be keen to know what new inventions will be impacting upon our lives 20 or 30 years hence, so we can be ready, even the experts often get it wrong. Forecasts commonly depend on projecting trends already evident forward by years or decades, but that cannot take account of the unexpected. How many of the inventions in this book would have been predicted to occur when they did? People still keep trying, using for example the Delphi Technique. This draws on the opinions of a number of knowledgeable people, but shares the ideas and then allows participants to change their minds, so that a consensus slowly emerges.

In 1964 the British magazine *New Scientist* asked dozens of informed practitioners how they thought their field of expertise would look 20 years ahead, in the much anticipated year 1984. Many of the commentators took a soft approach, not predicting any particular developments but rather outlining the problems they thought would be addressed over the next two decades and that might have found solutions, much as Francis Bacon did (p. 25) more than 400 years earlier. Others were more tough minded, and spoke with some certainty of what they expected to see. Some were

way off, perhaps because all the predictions were the work of individuals rather than groups, maybe reflecting personal passions and prejudices.

Others proved remarkably close to the mark. In the field of computers, the three or four commentators had similar visions. One expected the rapid progress of the previous 20 years to continue, leading to what he claimed, almost humourously, customers were expecting: 'a do-it-all, instant machine that operates at no cost and requires no programming'. Yet by 1984, the personal computers produced by Apple, IBM and others were beginning to meet that description (p. 285). They were indeed versatile, worked remarkably quickly, cost very little to run and came with pre-packaged software.

Another commentator referred to computers being 'omnipresent' in 1984, 100 to 1000 times faster than in 1964 but taking up only 1 per cent of the space, of international networks of computers, enabling information to be rapidly retrieved wherever it was held, of computers able to defeat the best human chess players. These predictions really did come true, and were even surpassed, if not by 1984 then in the decade following. Other equally firm predictions about the future are yet to be realised, even a further 20 years on. Computers have not in general abolished paperwork, or shortened our working week to four days. Everyday computers are not yet able to 'learn from their experience', still requiring detailed instructions through programming.

As for telecommunications, 20 years of progress would see (at least in the view of one commentator) much use made of 'coherent' light generated by lasers to carry information. That was indeed taking shape by 1984, though not in the way predicted. He spoke of light beams guided by lenses and mirrors through buried tubes. There was no mention of the technology that did emerge (p. 262), even though it was under development in 1964. Communication using orbiting satellites, just getting going in 1964, was expected to play an important role by 1984, as indeed it did (p. 257). But the anticipated 'video phones' did not proliferate, partly for technical reasons, and perhaps partly because users preferred the relative anonymity of voice-only communication. The increasing melding of the mobile phone with the digital camera may be about to change all that, more than 20 years after 1984.

Energy production was also prognosticated, with much emphasis on fuel cells (p. 252), which one expert believed would be very widespread in 1984, supplying homes and factories by efficiently converting methane or hydrogen to electricity. But unexpected technological problems have delayed the fulfilment of this vision for more than 20 years beyond 1984. Even today, fuel cells provide only a small fraction of our electricity needs. Another technology of which much was expected was MHD (magneto-hydrodynamics), which would generate electricity in large amounts with great efficiency by pushing very hot, electrically charged gases through strong magnetic fields. Again, there has been much difficulty in finding materials able to endure the hostile conditions and intense stresses.

Perhaps the most optimistic forecasts were made by those pursuing the 'conquest of space' (p. 174), five years before even the first steps on the Moon. They predicted that by 1984 manned spacecraft would have flown by the nearby planets Mars and Venus, and may even have landed on Mars. The latter is now not expected until 2020. Other predictions, nearer to home, were only a decade or so late, such as global networks of satellites to monitor the weather, or to provide navigational fixes for planes and ships (p. 277).

Optic Fibres Transform Communications

Charles Kao, George Hockham and Others

Optic fibres have been used in communications for only a few decades, but their power and potential are already evident. Trapping a beam of light within a tube by 'total internal reflection' is a familiar phenomenon. Jets of water, illuminated from within, have long been a feature of elaborate public fountains. By the early twentieth century, glass fibres were directing light into hard-to-reach places and, more importantly, allowing doctors to see inside the human body, for example into the stomach using a 'gastroscope'. Using light to communicate was not a new idea either. In ancient times sunlight reflected off mirrors could send coded messages, and Alexander Bell, co-inventor of the telephone, had his photophone (p. 133), though the light beam travelled through the air.

The potential for optic fibres in communications, especially when allied to the newly invented laser (p. 254), was clearly recognised in 1965 by Charles Kao and George Hockham at the British company STC, though they were not the first. They also recognised the fundamental problem. Impurities in glass absorb some of the light passing through. Even over say 5 metres (16 feet), ordinary glass will cut the intensity of the light by a factor of a million. Only specially crafted glass, cleansed of impurities so it was at least

50 times clearer than regular glass, could do the task. In 1970 Corning Glass in the USA came up with the first approximation, impregnating ('doping') silica glass with titanium. These fibres, like all since, had their maximum transparency when carrying not visible light but infrared radiation (akin to heat).

Optic fibres soon became an attractive alternative to traditional copper wires or coaxial cables for sending information from place to place. The much higher frequency of light means that vastly more information can be loaded onto a light signal than onto an electric current. One optic fibre could replace a whole bundle of wires or cables. Fibres were also largely immune to interference. In particular, two fibres running side by side did not interfere with each other (there was no 'crosstalk').

Over longer distances, such as spanning the oceans, the inevitable weakening of the signal was still a challenge. It had to be amplified to get to the other end. Initially that meant converting the light signals into electrical currents, boosting them electronically, and then turning them back to light, a complex and expensive procedure. In 1986 David Payne at the University of Southhampton and Emmanuel Desuvive at the Bell Telephone Laboratories independently found that an optical fibre impregnated with the rare element erbium could act as a light amplifier. Two years later, TAT-8, the first transatlantic telephone cable using optic fibres, went into service. This was barely 30 years after TAT-1 made transatlantic phone calls possible at all (other than by static-prone radio).

A glass fibre may sound simple to make but optic fibre manufacture is complex and exacting. Three or four different layers with slightly different optical properties are formed inside a thread only several times as thick as a human hair. Optic fibres can also be made more cheaply from polymers (plastic materials) but these are not as transparent. Initially fibres carried only one 'light' beam at a time, but modern fibres carry eight or more simultaneously, using slightly different frequencies or 'colours', in 'wavelength division multiplexing'. Using this technology, a single fibre can carry 'terabits' of information (millions of millions of bits) per second, corresponding to many thousands of simultaneous telephone calls.

In the early twenty-first century, all parts of our planet are linked by networks of optic fibre cables, many of them running beneath the oceans, with unprecedented (and still expanding)

capacity to carry information as text, voice or images. This has contributed to an equally unprecedented, indeed astounding, reduction in costs. Phone calls have become hundreds of times cheaper in real terms. A phone call across the planet, which cost half a week's salary in the 1950s, is now all but free. The quality is as good as if you were calling across the street, free of the distortion that was common on long-distance phone calls only a few decades ago. As the advertising slogan said: 'no one is far from anyone anymore'. Optic fibre communications also made possible the Internet.

Advances in optic fibres, light amplifiers, lasers, light-emitting diodes and other devices have now spawned a whole new technology called 'photonics', analogous to electronics but using light rather than electrical currents. New materials called 'photonic crystals' have been developed. These have regularly arrayed elements ideal for manipulating and controlling light, similar to the naturally occurring mineral opal, which shows beautiful iridescent colours. Researchers in this new field are confident that what we now do with electronics will eventually be possible through photonics.

The Explosive Terrorists Love

Stanislav Brebera

In December 1988 a mere 300 grams of the explosive Semtex, moulded into a cassette tape recorder, detonated on board Pan Am Flight 103 over Lockerbie in Scotland. The plane crashed with great loss of life. This wanton act made Semtex notorious and it became the explosive of choice for terrorist groups around the world. Invented in 1966, Semtex takes its name from its inventor's home village in the Czech republic. Stanislav Brebera was always saddened, he later said, when his invention fell into the 'wrong hands'.

Semtex is now the best known of the 'plastic' explosives. Their appeal, both for legitimate uses in mining and construction, and for criminal acts, is threefold. They are powerful, they are very stable and safe and, like plasticine, they can be made into different shapes, hence their name. Much of their potency comes from the explosive now called RDX (from its wartime code-name Research Development Explosive) and more ominously as Cyclonite. Like all explosives (p. 127), RDX gains its destructive force from exceedingly rapid chemical reactions, breaking down to release hot gases travelling so fast it is impossible to out-run the explosion, whatever action movies might suggest.

RDX is not new. It was first created in the 1890s (but initially as a medicine), found to be explosive in the 1920s, and much used on both sides of World War II.

Like other plastic explosives, Semtex has its RDX mixed with other compounds called 'plasticisers' and 'binders' to make it soft and mouldable. It can be safely shaped for maximum blast in a particular direction, say to crack open a safe, or packed into cracks in a wall and detonated to bring the wall down. Semtex, originally designed to destroy landmines, has no smell. Sniffer dogs cannot detect it, hence its appeal to users seeking to avoid detection. After the Lockerbie disaster, Brebera added metal fragments and chemicals with distinctive odours to make Semtex easier to find during security screenings.

Some energy must be supplied to get the chemical reaction going. Semtex and other plastic explosives like C-4 contain additional chemicals called 'stabilisers', so the energy boost that sets it off has to be very swift, from say a blasting cap; otherwise the explosive simply burns quietly. US soldiers in Vietnam sometimes set pieces of C-4 alight to warm their food, getting into trouble only when they stamped out the embers, the shock sometimes setting the explosive off.

Cash When You Want It

Who Invented the ATM?

So many people claim to have invented the ATM (automatic teller machine) that it must be judged simply an idea whose time had come. The first proposals for a 'hole in the wall' go back to the late 1930s, but the inventor Luther Simjian never patented his 'Bankomatic' and never succeeded in convincing anyone there was really a demand, though Citibank did trial it.

Fast-forward to the 1960s and the idea surfaced again in several places. In 1967 John Shepherd Baron, who ran a company called De la Rue Instruments, built machines to automatically dispense cash, including one outside a branch of Barclays Bank in north London, and later spoke about the development to a conference of 2000 bankers in Florida. At much the same time, James Goodfellow, working with Smith Industries, devised an automatic cash dispenser with some new ideas, including a keypad to nominate the money needed. Both Shepherd Baron and Goodfellow were later honoured by the Queen for their services to banking.

At least two other people have a claim, both Americans: Don Wetzel of the company Product Planning, which made automatic baggage-handling equipment, and John D. White, who reputedly built his first ATM in 1973. The argument about who invented what when will probably never be settled to everyone's satisfaction. Perhaps we can say that the two Englishmen pioneered ATMs in Britain, and the two Americans got ATMs going on their side of the Atlantic. We can also say that Wetzel at least, and probably several of the others, generated the vision of banking without a teller from the frustration of waiting in a queue, and the equally common experience of needing money when the bank is closed. Why should it be like this, they thought. Why indeed?

ATMs were not an instant success. They took several years to catch on among wary consumers. None of the early ATMs had the continuous connection to a central computer we see today, and through that to bank accounts from which the money could be drawn. That linkage would later make ATMs yet another manifestation of the information revolution. Without a link to an account, banks were wary about granting access to cash from such machines. An early challenge was identification, and there Wetzel seems to have the edge. His machines were operated by ID cards with a magnetic stripe, much like today.

Looking at Liquid Crystals

George Heilmeier, James Fergason

The face of your watch, or the panel on your microwave or iPod or almost any other modern device, will show you the vital numbers or words in the form of a liquid-crystal display (LCD), though some older ones might use light-emitting diodes (LEDs) (p. 256). These innovations have transformed the look of much of our technology.

Liquid crystals are an odd state of matter, not quite liquid, not quite solid. The molecules line up neatly as they do in a crystal, but they can move about as in a liquid, under influences like temperature, pressure or electric currents. First observed by an Austrian biologist in 1888, liquid crystals remained a laboratory curiosity until 1969.

George Heilmeier, from the American electronics company RCA, wanted a way to show TV pictures that did not rely on bulky, fragile cathode ray tubes (p. 161). By applying an electric current to a liquid crystal he could so affect the arrangement of the molecules that they would block light rather than let it pass through, or scatter rather than absorb it. So the crystals would appear light or dark on command. An array of them could display patterns, such as a TV picture. However, the initial devices were too heavy and inefficient to be useful. So he and his team turned to simpler problems, such as displays on watches and clocks. Heilmeier went on to head up the Advanced Research Projects Agency (ARPA), where the forerunner of the Internet had been devised (p. 269).

Better forms of LCD, using a different type of crystal and requiring much less power, were developed by American James Fergason. His initial interest was in measuring temperatures. Liquid crystals respond to temperature, and reflect colours preferentially depending on their state, so Fergason soon had a 'colour thermometer', placed, say, on a patient's forehead to display body temperature. He expanded into alphanumeric displays; watches using his LCDs ran for two years on a single battery, rather than merely two weeks.

LCD displays are ubiquitous nowadays. Laptop computers and even many desktops now use them, exploiting their flat shape. They have beaten off their early rival for displays, the LEDs, though LEDs have found many other uses. LCDs are now a multi-billion-dollar global industry. Three decades down the track even the first target, the LCD TV screen, has come to pass.

A Computer on a Chip

Ted Hoff, Stan Mazor

In 1969 the newly created microelectronics firm Intel, holder of the patent on the equally new integrated circuit (IC) (p. 249), was asked by Japanese manufacturer Busicom to make a set of integrated circuits to power a proposed line of programmable hand calculators. The job was given to employee number 12, bright Stanford graduate Marcian Hoff, known as Ted. Twelve was also the number of ICs to be created, each to do a different job: to receive the information, to do the calculations, to control the display screen and so on. Hoff argued that designing so many different chips would push up the cost of the device and computers in general; he decided to combine some of the functions, reducing the 'chip-set' to just four.

The most important of these would be a general processing or logic chip, able to do a variety of calculations and control tasks as requested (or programmed). With the help of Stan Mazor, he designed what would soon be dubbed the 'microprocessor' or 'central processing unit' (CPU). Others at Intel worked on ways to convert that design into the intricate patterns of silicon and other materials in a working microchip. The now legendary 4004 chip was launched by Intel in 1971. It held over 2000 transistors, and by itself was as powerful as the entire ENIAC computer of only two decades earlier (p. 224). It was literally a computer on a chip. The invention changed not only computing, still in its infancy, but the whole industrial world. Intel co-founder Gordon Moore, of Moore's Law (p. 282) called it 'one of the most revolutionary products in the history of mankind'.

Intel was not alone in this field, Texas Instruments (TI) was also active. Indeed Gary Bourne of TI secured the first patent for the single-chip microprocessor architecture in 1973. By 1981 Intel, barely 10 years old, had 20 000 employees, making nearly $200 million a year from microchips. Its success had spurred competition, often from firms like itself, set up by ambitious young engineers leaving more established companies. These were the boom times in Silicon Valley. By the mid-1970s, 50 or more types of CPU were on sale and their cost was down to a few dollars as the saturated market led to price wars. Microchips, both CPUs and RAMs, were now finding homes not only in calculators but in the earliest personal computers, arcade games and even home video games.

Within a few years Intel replaced the 4004 with the 8008, which could handle data eight bits at a time rather than four, and so was markedly faster in its operations. The 8008 was the ancestor of a dynasty of CPUs made by Intel, right through to today's Pentiums. Since that time, the progress of microprocessors has paralleled the other type of integrated circuit, the 'memory chip' (p. 282), and for the same reason. Advancing production technology has crowded ever more transistors and other components onto a fragment of silicon a few millimetres square. The number of transistors per chip soon became hundred of thousands, and today is hundreds of millions, as the typical size of components fell from one-tenth the thickness of a human hair to something a hundred times thinner again. Processing speeds have risen a thousand-fold to billions of cycles per second (gigahertz), though a high 'clock speed' is no longer the sole reason for a chip's power; modern chips take data in 64 bits at a time.

A major innovation, coming to light in the 1980s, was Reduced Instruction Set Computing (RISC). With increasingly sophisticated architecture, processing chips accept much simpler instructions. This made programming easier and increased competing speeds yet again. Most modern microprocessors use RISC or something like it. Designs are now so complex and sophisticated that humans cannot design them without the help of computers. The immense expenditure, amounting to hundreds of millions of dollars, required to design and test a next-generation microprocessor, and to build the high precision equipment ('foundries' or 'fabs') to make it by the million, has seen the smaller firms drop out. The mass production of microchips is dominated by a handful of large companies such as Intel, IBM and Motorola.

Microprocessors are not only unbelievably swift and sure, but they are also ubiquitous. Almost all machines or devices, from car ignitions to air conditioners and electric irons, from the office photocopiers and PABXs to your iPod and PDA (personal digital assistant), from traffic lights to industrial robots to probes in deep space, are under (mostly unseen) microprocessor control. CPUs assess the situation constantly and initiate action as needed in accordance with programmed instructions, much as a human would, but faster and more reliably, without losing concentration or needing a day off. Where will it all end?

A Foretaste
of the Internet

ARPA, Ray Tomlinson

One of the enduring myths about ARPANET, which was like a trial run of the Internet 10 years earlier, claims it was devised so the US military command and control system could survive a nuclear attack. If the large computers becoming so important for the national defence could be linked together to exchange information, the loss of one or more from a sudden nuclear strike would not mean the information was lost, and the whole system would not go down. While this is a reasonable motivation, those who were there at that time generally agree that this justification came later. There was a more immediate need.

The Advanced Research Projects Agency (ARPA) encompassed (and paid for) researchers across the USA involved in projects of military significance. They all could benefit from access to the large computers of the day, but those were relatively few in number and widely separated geographically. Putting these computers 'online', as we would say today, so that anyone could access them from anywhere, seemed a good idea, so it was done. The first computers, four of them in all, and all involved in research rather than operations, were hooked up to ARPANET in 1969. They connected Los Angeles, Santa Barbara, Stanford and the University of Utah through 56 kilobit per second links—top speed for the time. Through the 1970s, ARPANET linked computers in research labs supported by the Department of Defence across the country.

IP and FTP

To make the network operate, and to allow different sorts of computers to talk to one another and exchange information, the engineers devised some vital new technology, including the rules we now know as the Internet Protocol (IP) and the File Transfer Protocol (FTP). Central to these was 'packet switching', a vital part of the Internet's operations today. A piece of information, such as a file, passing along the network, was initially broken up into many small messages or 'packets', all the same size, like electronic envelopes. Each of them was given an identification number, showing where it came in the document, and a destination, before being sent on its way.

The network was studded with special computers known as 'switches'. These could read the destination of each packet and point it down the right wire. Whatever path the individual packets might take across the network (and the Internet nowadays has thousands of possible routes), they would all ultimately (within a few seconds) arrive at their destination and be reassembled into the original document. This was the beginning of something really big.

Email

ARPANET generated another major application, one about which people have mixed feelings. In 1971 Ray Tomlinson pioneered 'email'. A form of email already existed. People using the same computer could leave messages for one another. Each had an electronic mailbox that only they could open, but in which others could leave messages. A number of computers were already talking to one another through ARPANET. Why could not a message be sent to someone on another computer?

Once Tomlinson had thought of it, devising a new protocol to make it happen was easy. He simply piggybacked on an existing program that allowed computers on the network to exchange files. Tomlinson initially sent messages from one computer to another standing right beside it, though connected only through the network. Legend has it that his first messages were something like QWERTYUIOP, generated by simply dragging a finger over the top line of the keyboard. This is not of the same impressive tone as Alexander Bell's 'Mr Watson, come here, I need to see you' (p. 133), or Samuel Morse's 'What hath God wrought' (p. 104), but the impact on our daily lives has been at least as great.

Tomlinson's contribution to computer speak was the @ symbol. He needed a way to distinguish between messages going to people using the same computer, and those going out onto the network to someone somewhere else. By his own account he settled on the @, meaning 'at', since it was not used in anybody's name and could be taken to designate a place. It fitted comfortably between the user's log-on name and the name of their home computer. This terminology, so widely used today, was, he later admitted, the result of only half a minute's thought.

Tomlinson would also later admit that no one was clamouring for email at the time he invented it. It just seemed a 'neat idea', something that could be done. Yet few computer applications have had the impact of email, especially once it was married to the Internet (p. 290). Some people hate it, but others would confess they could not run their businesses, or even their personal lives, any other way.

Inside Insights:
CAT Scans and MRIs

Geoffrey Hounsfield and Others

Two inventions in the 1970s and early 1980s greatly increased the capacity of doctors and researchers to look inside the human body from without. One of them built upon an existing technology, the other found medical uses for a recent scientific discovery. Together, CAT scans and MRI scans have transformed much of medical diagnosis and therapy, and proved especially valuable in working out what is going on in a patient's brain and nervous system. Both inventions depended greatly on the development of high-speed electronic computers; both won Nobel Prizes for their inventors.

The discovery of X-rays late in the nineteenth century (p. 156) and their rapid application to medical issues like imaging broken bones brought immense benefits. Those were compounded 80 years later when British engineer Geoffrey Hounsfield first used the rapidly evolving computer technology to integrate a number of X-ray views of the body, initially taken one after the other from different angles. The usual X-ray film was replaced by crystal detectors and electronic amplifiers. The resulting

3-D image is usually presented as a series of slices across the body, from top to toe if need be, each slice being about 1 centimetre thick. The images are also much sharper and more detailed than traditional X-rays, able, for example, to pinpoint small tumours so they can be precisely attacked with radiotherapy. MRI was first in the news in 1972, when Hounsfield's machine located a tumour in the brain of a living, conscious patient.

C(A)T stands for computerised (axial) tomography, where a tomogram is an image of the internal structure of some object.

While Hounsfield had the first real success with a CAT scan, others had already been active. These included South African-born American physicist Allan Cormack, who was to share the Nobel Prize with Hounsfield in 1979, and American-born William Oldendorf who did not, despite writing many papers to promote the idea and securing the first patents. You can argue that British pop group the Beatles played a major role in the development of CAT scans; the massive profits from their records allowed electronic and music publishing firm EMI, where Hounsfield worked, to fund the research needed to perfect the CAT scanner.

Magnetic resonance imaging (MRI) works on very different principles. It was not initially used in medicine, but to study various chemical compounds. Its 1930s discovery won the Nobel Prize for Physics. Instead of sending beams of potentially damaging radiation from the outside, MRI effectively causes the body to scan itself from within by inducing water molecules to give out radio signals, revealing where and how closely packed they are. The human body is two-thirds water by weight. Different water content in various organs and soft tissues makes them show up distinctively on the scan; bones, which have little water, show up black. Diseases in organs are revealed by differences in water content.

To make all this happen, the patient must be subjected to a powerful magnetic field, more than 10 000 times as strong as the Earth's field. This does not in general pose any threat, other than, say, to people with pacemakers (p. 245). The field flips the atoms of hydrogen (strictly speaking, their nuclei) into new orientations, and they give off the distinctive radio signal when they 'relax to their old state. MRI has spread rapidly since scanners went on the market in the early 1980s. By the early twenty-first century, more than 20 000 of these expensive machines were in use and more than 60 million scans had been completed.

It took twice as log for MRI to be recognised with a Nobel Prize as for CAT scans. This may have been partly due to ongoing controversy as to who the actual inventor was. Some claim that American doctor Raymond Damadian should have shared some of the prize for his early work showing that cancerous tissues and normal tissues respond differently to the MRI technique. He certainly secured the first patents, and built a whole-body MRI scanner in 1977. Early histories of MRI attributed many of the early milestones to him, but later assessments claimed his technique had not been proven to reveal cancers.

When the prize was finally awarded in 2003, it went to American Paul Lauterbur and Briton Peter Mansfield: they had found how to process the signals with a computer to create 2-D and 3-D images, something that Damadian had not done. Yet many in the field, including the man himself, thought he had played a significant role in the creation of MRI scans, and ought to have been acknowledged. In modern science and technology, significant advances are often the work of more than one mind and heart.

Biotechnology Begins

Herbert Boyer, Stanley Cohen

Unlike earlier technologies created mostly by ingenious craftspeople, late twentieth century technology was increasingly driven by science. The massive biotechnology industry could never have emerged without deep understanding of the nature and operation of genes and the genetic code built up over decades of scientific research.

Many would name as the greatest scientific discovery of the twentieth century (perhaps of all time) the 'cracking of the genetic code' in 1953 by American James Watson and Briton Francis Crick, working together at the Cavendish laboratory in Cambridge, UK. Certainly finding a structure for the crucial genetic chemical DNA was profoundly important; in the wake of this discovery we have deciphered the complete genetic instructions for making whole organisms, including ourselves, and created the powerful tools and techniques that are now grouped together as 'biotechnology'.

Reading the code constituted a genuine breakthrough in biology. Now we could link the DNA in an organism's genes, including ours, with the particular protein each gene helped manufacture, and then to the body structures and processes the proteins built up or controlled. A change in a gene, say by a 'mutation' caused by radiation or some chemical, would affect the production of the protein the gene was responsible for and perhaps cause disease or disability. 'Genetic counselling' (advising people how best to manage an inherited or genetic condition), and even 'gene therapy' (replacing defective genes with properly functioning ones) now became real possibilities.

New knowledge does not inevitably lead to technology. Someone must take the first steps towards using the knowledge for a practical end, an end for which a demand exists or can be created. In the case of 'genetic engineering', a core element of modern biotechnology, the key date was probably 1972, and the key names, among many others, Americans Herbert Boyer and Stanley Cohen.

By then two important pieces of information had been added to the Watson–Crick revelation about the role of DNA. In bacteria, the simplest living things, some of the DNA is in small rings called plasmids. We also knew that DNA could be cut up into short pieces by special proteins called enzymes that always cut at the same point; other enzymes would join the pieces together again.

Boyer, Cohen and their colleagues extracted plasmids from bacteria and cut them open. They then inserted pieces of DNA taken from other organisms, rejoining the loose ends, to see if the bacteria would follow those new instructions as well as they did the ones they usually obeyed. They did; the bacteria began to manufacture the proteins coded for by the 'spliced-in' genes. Thus in 1972 began 'recombinant DNA technology', aka genetic engineering (though many now prefer the softer-edged 'genetic modification').

One practical consequence followed within a decade. Colonies of bacteria were making human insulin following instructions given by human DNA spliced into plasmids. By 1982 government authorities in the USA had approved the use of this insulin to treat diabetics in place of the animal insulin used until that time. Since then, hundreds, perhaps thousands, of chemicals for diagnosis or therapy have been created by genetic engineering. Much of medical practice has been transformed. By the 1990s the first genetically modified (GM) foods were coming onto the market, sparking a continuing controversy (p. 292).

The new technology has begun to repay its debt to science by helping knowledge grow. By splicing various bits of DNA into bacteria, including whole genes, we can see what the proteins they are coded for do, or if they do nothing. Early evidence suggested that many bits of DNA are 'junk', with no role at all.

This path to knowledge has not been unimpeded. Concern arose early that the most commonly used bacterium for this work (because it was so well understood) was *E. coli*, found in the human intestine, where it helps with digestion. If, say, cancer-causing genes were spliced into *E. coli* to see how they work, they could possibly have 'escaped' from the laboratory and infected the general population. So in the mid-1970s, researchers obeyed a call by the American Paul Berg for a moratorium on such experiments until proper safety measures were in place, including the use of variants of *E. coli* so weakened that they could not survive outside the test-tube. Once those safeguards were developed and legislated for, the work resumed and continues to the present day.

Not all biotechnology is modern. The making of bread, wine, cheese, yoghurt, beer and smallgoods, as well as the large-scale production of alcohol, are all examples of 'old' biotechnology, using living microorganisms such as yeasts and bacteria to carry out useful tasks, mostly through the process of fermentation. Much waste disposal, including sewage treatment, involves bacteria and biotechnology. Genetic engineering enables the microorganisms involved to be enhanced, fine tuned to deliver the best results.

Talking While Walking

Martin Cooper

On 3 April 1973, Martin Cooper, employed by electronic firm Motorola, made a call while walking in a New York street. He was using a prototype of the mobile or 'cellular' phone, which Cooper, as much as anyone, could claim to have invented. The telephone was the size of a brick and weighed a kilogram, far short of today's mobiles that weigh 100 grams or less and able to slip into a pocket.

Person-to-person communication using individual networks of mobile radios had been used for decades by emergency services like police and ambulances, taxis and long-distance truck drivers. Motorola supplied much of the equipment. Mobile phones would be different, connecting through the existing telephone system, which then linked callers by wire. Everyday phones were hardwired, 'landlines' in today's terminology. The emphasis would move away from the place and onto the person. Before the mobile phone arrived we would see the now barely remembered portable pager, able to receive a small message by radio (such as 'ring your office') but not send one. In time, the mobile phone would let the pager talk back.

Cooper's phone was experimental; a decade passed before mobile phones went on sale. The weight was down to half a kilogram but the cost was $3500. Uptake was understandably slow. The USA did not have one million mobile phone subscribers until 1990, though by then growth was becoming exponential as costs fell. The early networks were creaking under the load. Today mobiles number in the billions globally, with more mobiles than landlines now connected.

The concept of the cellular phone went back at least to 1947, the year of invention of the transistor, which would make so much difference to the size, cost and power of mobiles. The company AT&T (aka Bell Telephone) was an early mover (Cooper's call in 1973 was to an executive of AT&T, then their rivals in the business). The proposed system would operate by dividing the operating area into small zones called cells, each with a radio base station connecting to mobiles operating within the cell and in touch with other cells. As a user moved from one cell to another, the system would adjust automatically. Transmitting and receiving power could be kept low and the available radio frequency channels used more efficiently.

Unfortunately, the government regulator, the Federal Communication Commission (FCC), allocated so few channels to the initiative that hardly anyone thought it had a future; there was little incentive for research and progress was slow. Not until 1968 did the FCC reconsider its position; if the technology could be proven, more frequencies would be allocated. The chase was now on, with AT&T and Motorola in the lead. By the late 1970s trials of the new technology were running in cities like Chicago, with a few thousand handsets. A decade later the mobile was fast becoming mainstream.

Since then, consumer demand has joined with technological progress to make the mobile phone one of the most convenient and widely used inventions ever. Handsets have dwindled in size and weight due to more powerful computer chips and smaller batteries. More phones permitted smaller cells and reduced the required power to transmit messages. 'Analog' operation, like the old landline phones with the electric current varying in synchrony with the speech sounds, was replaced by digital, with the sounds transmitted in binary code, resulting much better reception. Successive generations of technology (many networks today use the 'third generation' or 3G) have used higher operating frequencies, carrying much more information. 3G phones can handle moving images as well as high-quality speech. 4G phones, not far away, will do even better. It has been a remarkable 20 years.

The other trend is 'convergence', with one device performing functions that previously needed separate machines. Most new phones include a digital camera (p. 279) and/or an FM radio. Many are almost pocket computers, storing data, sending and receiving emails, playing video games, accessing the Internet. They are already absorbing the MP3 player function, and soon will include a GPS device (p. 277) so users will always know where they are.

Yet the cost of buying them remains low and stable, even as functions are added. Phone networks even give them away. The capital costs reflect better technology and rising demand, with economies of scale. Most importantly mobile phones provide convenient access to data networks like the phone system and the Internet. That is where real value lies for the user, and how the supplier secures their profit.

Commentators may argue about the balance of impacts of the 'anywhere, anytime' capacity for communication that mobile phones bring. But that impact has been profound, perhaps greater than from any invention since the personal freedom of movement provided by the automobile.

Where in the World am I?

Ivan Getting

The latest 'must-have' for your car is a talking roadmap, sitting on your dashboard. From time to time a gentle voice tells you what turn to take to reach your destination. This would be impossible without the Global Positioning System (GPS), another achievement of the Space Age. The godfather of GPS was Ivan Getting, a research engineer at Raytheon Corporation, which had been very active in pioneering wartime radar and the first microwave ovens (p. 205). The first customer was the US military. A new generation of guided missiles was to be fired from railway wagons to achieve security from attack through being moved about. Their handlers needed to know precisely where their weapon was at the time of launch so it could be directed accurately to its target.

Leveraging ideas first developed during World War II, Getting proposed a network of radio transmitters, each with a precise clock. A receiver at the missile launch site, with a similar clock, would pick up several transmissions simultaneously. By noting how long each signal took to arrive, the missiles' minders could calculate the missiles' distance from each of the transmitters and hence their location.

In today's GPS, the radio transmitters are circling the Earth 20 000 kilometres up in precisely known orbits. More than 20 such satellites, each the size of a small car, are aloft at any one time. A GPS receiver (now only as big as a mobile phone and costing a few hundred dollars, thanks to microelectronics) uses transmissions from three or four satellites to calculate its position in three dimensions (latitude, longitude and altitude), accurate to a few metres. It is an astounding capability.

The USA launched the first GPS satellites in 1974. For a decade, only their military could use them; civilians gained access once the system was fully operational, in the 1990s. Defence authorities now access an enhanced system that reputedly provides positions to the nearest centimetre, and is used to guide 'smart' bombs and missiles. The systems cost about $6 billion to develop and nearly $1 billion a year to operate and maintain. However, economic benefits are piling up, especially from greater accuracy and safety of navigation of ships, planes and even land craft. One estimate sees savings reaching $50 billion a year by 2010. By then, GPS will have competition from systems launched by the European Union (*Galileo*) and the Russian Federation (*GLOSASS*).

The Supercomputer FLOPS

Seymour Cray

All computers have become much faster and more powerful over recent decades. But some are well ahead of the rest. A computer's speed can be measured in FLOPS (floating point operations per second), roughly the number of calculations done each second. The earliest computers (p. 224) did a few thousand, the first PCs (p. 285) a few million. Modern desktops manage a few billion, which sounds quick, yet leading edge 'supercomputers' are a thousand times faster again, and more. In 2006 a machine at a leading US defence laboratory completed an astounding 200 million million calculations each second (200 PetaFLOPS). Upgrades should take it to 500.

The pioneer was Seymour Cray, who worked for Control Data Corporation (CDC) in the 1960s, before moving out to start his own company. The first machine under this own name, the CRAY-1, reached the market in 1974. Building faster computers raised many issues. Cray figured that a number of smaller computers running side by side (aka 'parallel processing') would go faster than a single larger machine. To minimise the time taken to shift data from processor to processor, the connections were as short as possible. His most famous designs were cylindrical. All that processing in a small space generated a lot of heat. Lastly the data had to be moved in and out of the computer fast enough not to slow the calculations. As Cray said: 'Anyone can build a fast processor. What we need is a fast system'.

For several decades Cray-designed supercomputers (a term he did not use) whipped the opposition for speed. Nothing else came close, though CDC was nearly bankrupted building the early models. Cray hit financial turbulence himself later, perhaps not surprising in so risky a business, where one machine could cost $5 million. He was successful, selling 100 CRAY-1s and setting the early pace. Others have now taken over, using 'massively parallel' computing. Tens of thousands of processors like the CPU in your desktop run together to generate blinding speed.

The demand for speed comes from military agencies, universities and major research laboratories with big, tough calculations to do. Tasks include forecasting the weather (always a rugged one), figuring the shapes of chemical molecules used as medical drugs, simulating how new nuclear weapons or aircraft will perform (much cheaper than a wind tunnel) and cracking codes. By the way, only the top machines can beat the best human chess players.

Pictures with Pixels

Steven Sasson

The history of photography goes back about 200 years (p. 85). The many generations of cameras all had one thing in common: images were recorded on glass plates or plastic film using chemical emulsions that were sensitive to light. They were made visible ('developed') and permanent ('fixed') by chemical processing. That, of course, took time. Other than the briefly successful Polaroid process (p. 226), no camera took a picture you could look at immediately. Furthermore, the only way you could send a good-quality photo to someone was as a 'hard copy' (to use a modern term), printed on paper.

The comparison with today's 'electronic' or 'digital' cameras is startling. You can view the image as soon as you have snapped it, and you can send it down a phone line or cable, confident it will arrive looking good. And processing is easy. You can play with images, enlarge them, change the framing or colour balance, things that previously took hours in the 'dark room'.

This world of digital imaging is very new, for the everyday photographer at least. The first affordable electronic cameras reached the market only in the early 1990s. Their origins lie a few decades further back, and derive much from the development of television. That,

too, required technology to convert images into electric currents. In most digital cameras, the key component is the CCD (charged-coupled device), based, like the computer chip (p. 249), on semiconductors such as silicon. A CCD is an array of light-sensitive cells ('pixels') that accumulate electric charge in proportion to the brightness of the light that falls on them, much as solar cells do (p. 242). The level of charge on the pixels can be read out one by one, the data stored and processed to recreate the scene.

The first CCDs, built in the 1970s, were crude, with only 10 000 pixels (modern cameras boast millions of pixels). The images were correspondingly fuzzy. The first attempt to build one of these into a camera, undertaken by Steven Sasson at Eastman–Kodak in 1973, was equally clunky. The camera weighed 4 kilograms and took 20 seconds per image (black and white only). It was only a technical exercise to see if it could be done, not intended for the market. Kodak would not produce a consumer-friendly version for another 20 years.

But the exercise began an inevitable trend, one that will ultimately kill off the business that Kodak pioneered—everyday cameras using film. Such cameras are not yet dead, but they live on mostly in one niche, as the cheap, disposable 'cameras of last resort', used in emergencies or extreme environments such as underwater. These cameras trade off cost and disposability against image quality and just about every other feature of both traditional and digital photography. It is a rather sad farewell.

The evolution of the digital camera has been swift, in line with the rest of information technology. We see the same trends: ever-increasing capabilities at ever-decreasing cost. Early digital cameras cost $10 000 or more; that price has shrunk at least 20-fold. Progress has been aided by the development of international standards for storing images in digital code, such as the JPEG format devised by the Joint Photography Expert Group. Among other things, this allows images to be 'compressed'. A great deal of information in an image can be discarded without unacceptable loss of quality, as with DVDs (p. 293) and MP3 players (p. 291). As a result, images take up much less space in computer memories and can be quickly transmitted by email or over the Internet.

New sorts of data storage, such as 'flash memory', let a camera store more pictures. Inkjet printers generate quality copies on demand. Cameras continue to get smaller, now often barely larger than a credit card, able to be secreted in a mobile phone (p. 275) or in hard-to-reach places. Similar technology has transformed video cameras. Most electronic cameras will now take both still and moving pictures. We have come a very long way in little more than a decade.

The digital camera transformation raises a host of issues. Simplicity and ubiquity (such as in mobile phones) have encouraged some inappropriate use, with invasions of privacy that can verge on assault. The capacity to manipulate images by digitally altering the location and content of individual pixels has damaged our confidence that 'the camera never lies'. Surveillance in public places is easier and cheaper and therefore more common. This no doubt helps with the maintenance of law and order but asks a question about our right to privacy once we step out of our front doors. Digital cameras are one of the ways we can create and access information freely for ourselves, a capacity central to the information age, open to both use and abuse.

Making the Magic Bullets

Georges Kohler, César Milstein

The development of 'monoclonal antibodies' ('MABs') spanned the sometimes-unclear boundary between discovery and invention, between science and technology. Based on a better understanding of how our bodies work, it also required innovative techniques if that knowledge was to improve human health. Monoclonal antibodies give form to Paul Ehrlich's vision of 'magic bullets' (p. 183). If chemicals could be found or created that selectively targeted a disease-carrying organism, something to kill that organism could be delivered at the same time, without damaging surrounding tissues.

Our bodies already make one form of such bullets. Antibodies are manufactured by particular white blood cells, called B-cells, when we are invaded by bacteria or viruses. Some antibodies (thousands of varieties are known) recognise cells that have turned cancerous; others can target our own healthy tissues, misjudging them as foreign and causing 'auto-immune' diseases such as rheumatoid arthritis. Using antibodies deliberately for diagnosis and therapy needs large quantities made on demand in the laboratory. Medical researchers Georges Kohler and César Milstein, both working at Cambridge University, won the 1984 Nobel Prize for inventing a way to do this.

In 1975 they began with B-cells that had become cancerous (the condition is known as 'myeloma'). They were multiplying uncontrollably, becoming essentially immortal. The researchers fused the myeloma cells, which had lost the power to secrete antibodies, with other healthy B-cells that were producing the antibody they were interested in (each variety of B-cell produces only one pure antibody). The resulting hybrid cell (a 'hybridoma') divided continuously into a vast number of exact copies ('clones'), all making the same (monoclonal) antibody.

The initial work used mice, and MABs are still often made that way, though since the 1980s, hybridomas can contain human genes as well. This reduces side effects in human patients. Successful applications have been slow to emerge. The patient's immune system mobilises against antibodies from mice, limiting their effectiveness. Nonetheless, dozens of MABs have been approved since the mid-1980s to act against transplanted tissue rejection and auto-immune diseases (by targeting the immune system cells that cause them), blood clots following heart surgery (by targeting platelets) and a range of cancers (by locking onto some identifier on the cancer cells and stirring the rest of the immune system into action). There is still a long way to go to get the most out of MABs.

The March of the RAMs

Robert Dennard

Moore's Law is not so much an invention as a description of the impact certain inventions are having. It is not really a 'law' either. It does not state how information technology must behave but how it has behaved, and is likely to continue to behave.

Gordon Moore co-founded the microelectronics behemoth Intel in 1970 with Robert Noyce, both having left pioneering firm Fairchild Semiconductor. They planned to push the new integrated circuits or ICs (p. 249) as far as they could go, and that has proved a very long way indeed.

Moore had noted early that the number of circuit elements (transistors, capacitors, resistors) that could be squeezed onto a less-than-fingernail-sized piece of silicon (a 'chip') had doubled every year since the IC was invented. He first stated what someone else called 'Moore's Law' around 1965. And he saw no reason why this once-a-year doubling should not go on. More components on a chip would mean it could do more, or do the same faster. Computers looked set for unprecedented growth in power and speed, with simultaneous massive reductions in size, weight and cost. So it proved.

Consider the 'random access memory' (RAM) chips that store data digitally, ready for very fast access. These were invented in 1966 by Robert Dennard of IBM, with one transistor and one capacitor combining to store one 'bit' of data, a 0 or a 1. This was the first step on a very long road. Progress seemed quite slow at first but picked up as Moore's Law began to bite. By 1975, RAM chips held enough components to store 1000 bits (a kilobit or 1 Kb), sufficient to put the existing computer memories (based on magnetic cores) out of business.

Even a kilobit is not much data, equal to 125 letters or numbers, given that eight bits are needed to represent a single character. But progress was now relentless, though the doubling time over the long haul proved to be more like 18 months than 12. As microchip components shrank in size, the capacity of a typical commercial RAM reached 4 kilobits in 1978, and 16 kilobits in 1980. Sixteen kilobits corresponds to about a page of text. In 1983 came the 64-kilobit chip, used in early personal computers like the Commodore 64 (p. 285). The pattern was now obvious to anyone. The 250-kilobit chip duly arrived in 1985 and the 1000-kilobit or 1 megabit (1 Mb) chip entered the market pretty much on cue in 1988. RAM capacity had risen a thousand fold in the 13 years since 1975. Moore's Law had held good for more than two decades.

It would continue to hold. The milestones passed, and around 2002, after another thousand-fold increase since 1988 (and a million-fold increase over the RAM of 1975), we had the 1000 Mb chip (the gigabit or 1 Gb chip). This tiny device, with tens of millions of components in a few square millimetres but no moving parts, had enough memory to hold 10 minutes of MPEG-compressed movie (p. 293) or the information content of 250 books, every word instantly accessible.

Ultra-miniaturisation continues apace, even though the dimensions of chip components are now exceedingly fine, around one-hundredth the thickness of a human hair. Sooner or later the chipmakers will run into some fundamental barrier to do with the structure of matter, but that seems a decade or more off. If Moore's Law continues to hold, we will, say by 2015, be approaching the next milestone, a chip holding a thousand gigabits, a 'terabit' of data, the information content of 250 000 books or 200 hours of CD sound, or 30 movies. All other storage media may be facing obsolescence.

One other comparison shows how far the path charted by Moore's Law has taken us in five decades. In 1955 some millions of individual transistors were made, costing $5 each. Early in the twenty-first century, annual production was a billion times greater, with transistors clustered by the million on microchips, and the cost per transistor had fallen correspondingly to one twenty-millionth of a single cent. We had never seen a transformation remotely like it.

The real impact of these stunning statistics lies in the uses we make of the technology. With RAM chips to hold data and the equally formidable microprocessors (p. 267) to compute that data billions of times a second, our desktop computers do just about whatever we ask with words, figures, sounds and images. We can create, store, process, transmit and display information as, when and where we wish, easily, quickly, cheaply and simply. In addition, computers perform unseen work controlling appliances, vehicles and other machinery. But our appetites continue to grow. As the technology has advanced, we have continued to find uses for it. That too is likely to continue.

The Girl from the 'Test Tube'

Patrick Steptoe, Mu Chueh Chang, Robert Edwards, the Monash IVF team

Louise Brown has a unique place in the history of medicine. She was the first human child to be born after conception, not in her mother's body but in a piece of laboratory equipment. This 'test tube' baby', as the popular press named her, was the outcome of 'in vitro fertilisation' (IVF—meaning 'in glass'), though neither test tubes or glassware were involved. The birth was a triumph for British doctors Patrick Steptoe and Robert Edwards, but they had built on the achievements of others, and researchers elsewhere quickly enhanced their work.

Artificially creating an embryo by bringing male sperm and female eggs together outside the body, and subsequently implanting that embryo in a womb for natural growth and delivery, had been proven possible in rabbits in the 1950s by the Chinese–American reproductive biologist Mu Chueh Chang. Human eggs were successfully fertilised in the laboratory as early as 1973, but the embryos did not survive once implanted in the mother's womb. With improved techniques, success finally came in July 1978 and the birth of Louise Brown. Controversy over the safety, and indeed the ethics, of IVF soon followed and have not subsided.

Leadership in the new technology soon passed to a team at Monash University in Melbourne, Australia, involving among others Carl Wood and Alan Trounson. Their initial challenge had been to restore fertility in women whose fallopian tubes had been damaged by disease, preventing eggs released by the ovaries passing through to be fertilised. They failed in their efforts to replace the damaged tube with an artificial one and the emphasis turned to fertilisation outside the body. The Monash team devised a regimen of hormone therapy that would cause the ovaries to release more eggs than usual and in a controlled manner. Steptoe and Edwards had relied on the natural monthly cycle. The Monash program produced 12 of the first 15 IVF babies in the world, followed by 8000 more.

IVF has successfully addressed many fertility problems beyond fallopian tube damage, such as poor sperm quality and endometriosis. Available in most counties, IVF now initiates 1 per cent of all births globally. The proportion is up to 4 per cent in Demark. Research continues, seeking to improve the overall success rate, which is still around 20–30 per cent for a single course of treatment.

'The Computer Moves In'

The Rise of Personal Computing

Early electronic computers (p. 224) were massive, expensive and by today's standards ponderously slow, even if faster than hand calculators or slide rules (p. 26). New technology was beginning its march, with the use of transistors (p. 228), and a decade later integrated circuits (p. 249), hacking into computer size and weight. By the early 1970s the typical 'minicomputer' was only as big as a refrigerator or a grand piano, though still for collective rather than individual use. Anyway, who would want one for themselves?

Most people old enough to remember would say that the personal computer (PC), such as most of us now have on our desks and generally reserved for our use alone, entered our lives in the mid-1980s. That's true overall, but the early movers were already a decade down the track. In the early 1970s, the Xerox Alto, produced by the Xerox Corporation (p. 210) was the typical fridge size but becoming user-friendly, though very few people got to use one. It was the first computer with two features that later became almost universal: a 'mouse', invented by Douglas Engelbart in 1963 to move a pointer around a display screen, and later refined by others; and a 'graphical user interface', images and objects ('icons') on the screen at which the user could 'point and click' and so tell the computer what to do next.

The Apple II and The Commodore

The 1970s were the heyday of computer hobbyists, the original geeks, happy to assemble computers like the 1975 Altair from a pile of parts with a soldering iron, and to meet with other nerds to swap war stories. Among the people writing 'software' for the Altair were Paul Allen and his high-school friend Bill Gates, who later co-founded the leviathan Microsoft. Also active were the two Steves (Jobs and Wozniak), co-founders of Apple Computers. Their Apple II, released in 1977, quickly became the market leader, typifying the personal computer phenomenon then stirring. It had coloured graphics and the new 'floppy disc' drive, first devised in the 1960s but now cheap and reliable enough to replace cassette tapes as a way of loading programs and data.

It was innovative in software too, particularly in supporting the Visi-Calc spreadsheet (Dan Bricklin and Bob Frankston, 1979) for users needing to juggle figures. This 'killer application' made computers attractive to business, a major new market. Users today would think both processing speed (around 2 megahertz) and onboard memory (typically 4 kilobits or 100 words) absurd, but such computers found buyers then, with more than a million Apple IIs sold for $1000 each, a lot of money at the time.

Released in the same year as the Apple II were the Commodore PET and the TRS-80, museum pieces now but with many devotees in their prime. The PET's successor, the Commodore 64 (with an unprecedented 64 Kb of onboard memory) released in 1982, was to become the most popular single model ever; between 1982 and 1994 it sold 17 million units.

PC vs Mac

The giant firm IBM ('Big Blue') had been very successful in the 1950s and 1960s with 'mainframe' computers; these had a large central processing unit supporting many terminals. Now IBM was on the sidelines. It had seen no future in minicomputers and built none, but in 1980, spurred by the success of Apple, decided to enter the new 'microcomputer' market. The first IBM PC reached the stores in August 1981. Its impact, and that of Apple, was so great that *Time* magazine named the personal computer 'Person of the Year' for 1982, the first non-human so acknowledged (the US and Soviet presidents shared the honour the next year). The 3 January 1983 cover of *Time* announced with foresight 'the computer moves in'.

In 1984 Apple released the first Macintosh, which would be IBM's major competition into the early 1990s, but IBM 'clones', with the same structure as the IBM PC and using the same software but with a variety of manufacturers, were soon proliferating, the competition exterminating some of the pioneers such as Commodore.

Towards Today

Into the 1990s, new features and functionalities were emerging, 'hard drives' (p. 246) within computers to store data and programs for quick access and use, compact discs (p. 287) initially as CD-ROMs (read-only memory) that loaded programs faster, systems to deliver CD-quality sound and more realistic graphics for computer games. Software programs proliferated and became more sophisticated (and so needed more internal memory to store them), programs for word processing, databases, spreadsheets. Computers in offices and even in homes were now sending email (p. 269) and accessing the Internet (p. 290).

By 2000, most of the features of today's desktop computer were in place and the 'laptop' was a realistic description, with smaller, lighter batteries (p. 78) and more compact circuits. Personal computers were everywhere. In barely 20 years, they had become all but indispensable.

The Compact Disc Takes Over

David Gregg

The arrival of the compact disc or CD in the marketplace in 1983 was both a beginning and an end. It brought a profound change for music lovers. Audiophiles and classical music buffs embraced it at once for its increased fidelity and convenience; the rock and pop communities were somewhat slower to enthuse but ultimately did. At the same time, it signalled the close of almost a century of sound recording on vinyl discs, since the days of Edison and Berliner (p. 135). It heralded the end of analog recording, with CDs using the newer 'digital' techniques, and of the seemingly inevitable clicks and pops that plagued the old 78s and LPs. CDs seemed almost impervious to dust and damage.

The idea of a disc holding information that could be read by a beam of light rather than a needle seems to have come first to inventor David Gregg in the late 1950s. He wanted to record movies; his early optical disc was soon called a video disc, and then a laser disc when the laser (p. 254) became the ideal tool for both recording and playback. Gregg's patents (and his company) were acquired by the firm MCA which, in partnership with Dutch electronics giant Phillips, commercialised the laser disc for movies, reaching the market in 1978. This was 15 years ahead of the DVD, which would ultimately do the same job much better (p. 293). Laser discs never really caught on. They were inconveniently large (30 centimetres), played only for an hour (so holding only half a typical movie), and were not digital, with the limitations that brought.

Philips and others, including Japanese electronics leader Sony, were also experimenting with using laser-disc technology to record sound only. It seemed sensible for Philips and Sony to cooperate. Combining ideas and agreeing on standards, they created what we would now recognise as compact discs. These high-quality polycarbonate plastic discs were backed with reflective aluminium and protected by a lacquer coating on which labels could be printed.

The key development was going 'digital' (or more precisely going 'binary'—using only the digits 1 and 0). Rather than continuously recording every minute fluctuation in the sound waves conveying the music, engineers devised ways to take 'snapshots' of the sound wave, relying on the human ear and brain to fill in the gaps. It parallels the way a series of still images projected quickly enough can give the illusion of movement. A measurement of the strength of sound waves, taken 44 000 times a second, is converted into a binary number, usually one with 16 'bits' allowing for thousands of distinguishable levels of sound. For most of us, the ear itself cannot do better.

The processed music was laid down (and still is) in a spiral groove cut into the polycarbonate, as on the old vinyl records, but only about a micron (a thousandth of a millimetre) wide. The digitised data was encoded as changes in the depth of the groove, much as in the Edison phonographs, rather than the side-to-side wobbles popularised by Berliner. Just two distinguishable depths ('pits' and 'lands') accommodate the two-symbol digital code. On playback, a laser beam reads the pattern of highs and lows, turning the signal back into music.

CDs had great appeal from the start. Smaller and more convenient than the 30-centimetre LP, they can hold four times as much music; the target playing time of around 70 minutes was set, so the story goes, by Beethoven's Ninth Symphony. CDs are much less susceptible to damage, and some players can be carried about and even bumped without interrupting the music. In the judgement of most people, the sound quality is at least as good, though some traditional audiophiles will never accept that. Some claim, for example, that the dynamic range (from the loudest sound to the softest) is compressed compared to an LP.

Although CDs arose originally from a search for a way to record movies, and initially were used exclusively for music, they soon became a convenient and powerful way to store any information that could be put into binary code. That meant just about everything: text, numbers and still pictures were soon being recorded for playback from CDs. From the 1990s the CD-ROM (read-only memory) supplemented and then replaced the floppy drive in personal computers. Given its power and convenience, you would have thought the CD was here to stay, but it is already under challenge from the increasing capacity of microchips (p. 282), which have no moving parts and store large amounts of information. Hence the proliferation of iPods and other MP3 players (p. 291). In today's technology, nothing is forever.

The Ultimate Hearing Aid

Graeme Clark

Deafness can severely impede education, employment and enjoyment of life. Early aids to the deaf included the ear trumpet, which projected extra sound waves into the ear. Electronic hearing aids took over in the early twentieth century, but these were initially large and clumsy. Batteries and amplifiers were carried in a small suitcase; sound was collected with a microphone and sent amplified into an earpiece. With the advent of transistors, hearing aids shrank in size until they could be worn, even fitted behind the ear, and microprocessors allowed performance to be tailored to individual needs.

These aids sought to overpower the inefficiencies and losses that sound encounters as it makes its way though the ear towards the hearing organ. A different approach was promoted in Australia in the 1960s, the 'cochlear implant' or 'bionic ear', created by Melbourne-based hearing researcher Graeme Clark. This bypassed the ear, instead using small electric currents to directly stimulate nerves in the inner ear or cochlea, where impulses are generated to go to the brain. This could help the profoundly deaf whose hearing had been destroyed by severe infections such as meningitis or German measles in utero or during childhood.

Cochlear implants do not restore normal hearing. Instead they process the sound and direct impulses through implanted wires to 22 separate points in the cochlea. This makes the user aware of their sound environment and helps them understand speech, though that requires much training and post-implantation therapy. Since implants were first approved by the US Food and Drug Administration in 1985, 100 000 people, half of them children, have benefitted from the technology. Most are in developed countries; the device and necessary surgery are costly.

Clark was inspired by the problems suffered by his hearing-impaired father and by reading of work overseas to stimulate auditory nerves. He struggled for a decade to secure resources for research, finally securing support through a TV telethon in 1974. Government funding came later. The implant was commercialised by a medical equipment supplier later renamed Cochlear Ltd, which now supplies the bulk of implants worldwide.

The innovation has not been universally welcomed. Many of the deaf community who use 'signing' and lip-reading to communicate believe that restoring hearing through the implant endangers their traditional practices and culture.

The World Wide Web

Tim Berners-Lee

With hundreds of millions of people sitting at their computers and accessing the Internet everyday, connecting to other computers across the world, few would doubt its power and influence. But the Internet is just a computer network, or a network of computer networks. Its value comes from the way we use it, to send emails, for example, or to access the World Wide Web (WWW).

The web is a remarkable system, enabling anyone to quickly access a publicly available document or other information source wherever it might be, and for no cost other than the Internet connection, to read it, print it out and store a copy on their own computer. It is the most far reaching and powerful means for exchanging information that has ever existed on this planet.

For all its reputation, the web started small, initially involving just the scientists at one European research centre, the CERN particle physics laboratory near Geneva. One of those people was Tim Berners-Lee, from England. Many of the issues involved in sending documents through the Internet had already been solved through the workings of ARPANET in the USA (p. 269); the question now was how to find what you wanted to have sent to you.

In 1989 Berners-Lee proposed 'a global read–write information space' where each information source—a scientific paper for example—would have its own unique ID, a 'uniform source identifier'; programs on your computer called 'browsers' or 'search engines' could lead you to it. Within two years, the World Wide Web (his term) was in use across the particle physics community; within another two years, other academics and then business were taking notice. The latter soon realised that enabling customers to find you on the web (having a 'web presence') through their own webpage or website was more than useful; it was essential.

In 1994 CERN announced that anybody who could connect to the Internet could use the web, and it would be free. Not surprisingly, the following years were a period of extraordinary growth in web use and web presence, the latter including the new phenomenon of 'blogging'.

Berners-Lee's honours include a knighthood in 2004, the Fellowship of the Royal Society in 2001 and being named by *Time* magazine in 1999 as one the 100 most influential people in the twentieth century. Given the pervasive influence of the WWW, the last may understate the case.

More Music in Less Space

Music Compression Becomes Mainstream

Just about everyone knows the term MP3. They know it lets them replay dozens of popular songs from a machine the size of a deck of cards. MP3 players are transforming the convenient enjoyment of music. But the name has a deeper meaning. MP3 is shorthand for MPEG1 Audio Layer 3. MPEG means Moving Picture Expert Group, a group of boffins meeting since 1988 to set internationally accepted standards covering the encoding of video and audio signals (aka pictures and music) into forms that can be quickly moved electronically, say over the Internet.

The MP3 players draw together several technologies. Storing music as strings of 0s and 1s, MP3 is yet another manifestation of the digital revolution. The player holds these bits of information for quick retrieval. Unlike a CD player, an MP3 player has no moving parts; the data is imbedded in a microchip. The operation of Moore's Law over decades has produced chips able to hold immense amounts of information (p. 282).

A problem remains. Using the sampling methods used in compact discs (p. 287), a typical three-minute popular song needs 32 megabytes (32MB) of data to capture all its nuances. Downloading that from the Internet takes an inconvenient 20 minutes through a 56-kilobit 'dial-up' modem. To make music sharing through the Internet practical, this download time had to be reduced by 'compressing' the data needed to encode the song without making it sound noticeably different.

Hence MP3, the current answer to the problem as devised by the bright people working under MPEG, mostly European engineers funded through the European Union. Applying a few rules to remove redundancies, such as not coding sounds the ear cannot hear, and storing only the louder of two sounds when they sound together, a three-minute song can be stored in only 3MB, 90 per cent less space. This cuts the download time to two minutes, less time than the song takes, but keeps (just about) CD quality. With broadband, it's much faster still.

MP3, which followed industry standards MP1 and MP2, entered our lives in 1994, when the first programs to convert music to the MP3 format went on the market, and in 1995, when MP3 files began to flourish on the Internet, ready for download to portable players. Once more, life would never be the same again.

Flavr Savr Fails

The First Genetically Modified Food

In 1994 the US Food and Drug Administration (FDA) announced the first approval for sale of a genetically modified (GM) food. Genetic engineering (p. 273) had already created strains of bacteria producing human insulin or growth hormone, but not before had a plant been genetically altered. The Flavr Savr tomato addressed a longstanding marketing problem. Ripening tomatoes become softer as pectin in their cell walls breaks down naturally, and so are more easily damaged in packing and transport. Consequently, tomatoes are picked green and ripened by exposure to ethylene gas. The genetic tweaking in Flavr Savr slowed the pectin breakdown, keeping the tomatoes firm while they ripened on the vine and gained more flavour.

Behind the new tomato was the California-based biotechnology firm Calgene, one of the new breed of firms commercialising genetic engineering, as Silicon Valley had done with semi-conductors a generation before. Although the Flavr Savr tomato was FDA-approved and found to pose no identifiable risk to human health, it failed in the market place. Customers were divided about its flavour, there were problems ensuring supply, and competition from a conventionally bred tomato with the same appeal. There was also vocal opposition from the critics of GM, who derided the produce as 'frankenfood'. After a couple of years, the tomato was withdrawn from sale and Calgene taken over by the chemical giant Monsanto.

The demise of Flavr Savr was not the end of GM plants, though GM fruits and vegetables still struggle for acceptance. In one line of development, food crops like canola and soybeans are given resistance to common herbicides, simplifying weed control (you can spray the weeds without killing the crops). Monsanto has been active there. In another, crops like cotton and corn have been given the ability to make a toxin naturally found in a bacterium, and so to kill the insect pests that afflict them. This reduces the need to spray synthetic insecticides.

Another well-publicised initiative increases the iron and vitamin content of rice, producing 'golden rice' with substantial health benefits. Other GM crops have enhanced tolerance to heat or drought or to saline or acidic soils. At the same time, a lively debate continues about the safety and desirability of GM (or 'transgenic') crops and other organisms, their benefits compared with their costs. Government policies vary country to country. GM plants are probably here to stay, but it is not clear how far they will gain acceptance.

More Movies in Less Space

The DVD Triumphs

With CDs (p. 287) replacing venerable vinyl discs for music recording from the 1980s, a push began for an optical disc to record movies. For home viewing, movies had already been transferred to video cassettes (p. 243). Industry leaders remember well the costly battle between the Betamax and VHS formats; so IBM president Lou Gerstner acted as mediator between rival videodisc formats, one pushed by Phillips and Sony, the other by Toshiba. Phillips/Sony ultimately withdrew, but some special features they wanted were included in the victorious (and more heavily backed) Toshiba technology.

The compromise produced the DVD—the Digital Video Disc or Digital Versatile Disc (since they can record more than movies). DVD players and discs went on sale first in Japan in 1996, with other countries quickly following. VHS, dominant until then for home video entertainment, was doomed. In the USA, DVD rentals surpassed VHS hire in 2003; in a few more years VHS will disappear altogether, except for nostalgia.

Like a CD, a DVD records information in digital code as pits in the bottom of a spiral groove, burnt by a laser and read out by another. However DVDs can hold six times as much data, 4.5 gigabytes (GB) rather than 700 megabytes. The grooves are twice as fine, the pits half as long; more space is saved by efficient error correction procedures.

Even so, fitting a two-hour movie onto a 12-centimetre disc needs more innovation. Moving images contain immense amounts of information, but much of it is redundant. If a large area of blue sky on one frame does not vary significantly from place to place, and is the same in the next frame, it is enough to record that it has not changed. Other details can remain unrecorded because the eye cannot perceive them. Using the data compression technology MPEG2 (p. 291), the DVD can store 40 times less data than the printed film without the viewer noticing the difference.

The now-familiar DVD is not for the long term. Already (2006) rivals and successors are mustering. High-density DVDs use blue lasers rather than red ones to record and play back; the pits can be smaller and more data (up to 15GB) recorded. Those could still be outclassed by the rapid growth in data capacity of hard drives (p. 246) and even microchips (p. 282). Time will tell.

A Clone Called Dolly

Hans Spemann, Ian Wilmut

The idea of a 'clone', an individual created to be identical in every aspect to another, always sounded like science fiction. Of course, nature achieves it whenever identical twins are formed, each half of a dividing embryo becoming a fully formed person. Perhaps the first to do this artificially was the German researcher Hans Spemann. In 1902 he divided the embryo of a salamander (a reptile similar to a lizard) containing just two cells, using as a knife a hair from the head of his infant son. Once separated, both cells developed into complete, genetically identical animals. In another experiment, he removed one cell from a 16-cell embryo; both the large and small embryos became complete functioning salamanders.

Spemann later proposed a 'fantastical experiment': to take genetic material from an adult cell and use it to create another identical creature. Though Spemann never managed it, his process, now called 'nuclear transfer', is the basis of cloning today. Success first came in 1951, when US researchers cloned a frog, inserting the nucleus from an embryo cell into an unfertilised egg. Experiments with mammals were more difficult because mammal egg cells are very small. Claims emerged of cloned embryos from mice, cows and sheep but no one produced a viable living creature. Some of the experiments were later found to be fraudulent.

In 1996, after a decade of work and nearly 300 trials, Scottish researcher Ian Wilmut cloned a sheep, named Dolly after the US singer Dolly Parton, by transferring the nucleus of an adult mammary cell also into an unfertilised egg. The embryo so formed was implanted in another sheep's womb to grow until birth. It was later discovered that because her 'parent' was six years old, so were Dolly's cells when she was born. Her life expectancy was therefore less. A year later researchers in Hawaii cloned a mouse and continued the process for three generations, producing 50 clones. Japanese researchers produced eight identical copies of a cow. By 2000 pigs and rhesus monkeys were being cloned.

Given the work done to date, cloning humans ought to be possible, but most countries have banned it on legal and moral grounds. Researchers look to cloning of other animals to produce strains with desirable characteristics, such as producing milk with medically beneficial qualities. Cloning will increase our understanding of genetics and reproduction, and may lead to the re-creation of extinct species such as the mammoth.

Where To From Here?

This book has charted the progress of invention and innovation in Western societies over the last 500 years. It describes the circumstances under which important new technologies have arisen, and the impact they have made—some of them still vital today, many more displaced by the newer and better. In looking forward to what new wonders may arise in coming years, the best available guidance comes from what has already happened or is happening now.

Technological change is driven by a combination of factors: consumer demand, the need for corporations to make profits, and social needs in areas like health and defence. It has strong momentum, and we can assume that, baring some catastrophe, existing trends will continue into the foreseeable future, though new factors that we do not yet know about or cannot predict will speed up progress or slow it down. We can do as Francis Bacon did and make a wish list of advances we would like to see. If the need is strong enough (consider the project to develop the atomic bomb) people and resources will be diverted to the new lines of research and development. But most technological development proceeds along well-laid tracks.

This means that new and better solutions to our needs will continue to emerge. 'Better' can mean many things: more compact, safer, more convenient or user-friendly, more efficient, using fewer resources, less polluting. Based on previous experience, we know costs will continue to fall. What is too expensive for everyday use at one time will become commonplace a few decades later. We have seen this already in telephones, automobiles, television sets, household computers and international air travel. The benefits of technology are now more equitably spread through society than they have ever been.

Think how many of the devices and systems we now take for granted emerged from the laboratories onto the market only in the last 20–25 years—less than a generation. The list includes CDs, DVDs and MP3 players, personal computers with all their software, video games, email and the Internet, mobile phones, still and video digital cameras, 'personal digital assistants', automatic telephone exchanges in offices, fabrics like Gortex and Spandex, genetic engineering, CT scans and MRIs, ATMs and EFTPOS. Most of these are information technologies.

This may prove to have been an abnormally productive period, a unique meeting of new ideas and a receptive market unmatched in the history of innovation, but

I would not count on it. The pace of major, life-changing innovation may not accelerate. The market may have a limited capacity to accept an endless stream of new gadgets and functions. But it is unlikely to slow. New things will continue to arrive; many are already approaching.

What then of, say, the next decade? What are we likely (or unlikely) to witness? Based on what is already in the works, here are a few predictions.

Information Technology

Information technology (IT) will deepen its impact on every aspect of our lives. Convergence, already typified by the proliferating functions of a mobile phone, will continue to be the dominant trend. There will be fewer new devices but more functions in the ones we have. Through its access to the Internet (or its successor), a single hand-held machine or (or even a wearable one) will provide mobile phone services, Internet and email access, computer functions like word processing and spreadsheets and substantial data storage. It will generate still and video images, replay music and video on demand, tell you where you are and how to get somewhere else. We have most of that already. The number of Internet users will pass one billion, with most connected by broadband.

Moore's Law, or something very like it, will continue to hold for at least the next decade. The complexity of microelectronic devices will double every few years or less. This will come from the ongoing microminiaturisation (or now perhaps now the 'nanominiaturisation') of processors and memories that crams ever more components into the same space. The typical dimensions of chip components have fallen from about 1000 nanometres (millionths of a millimetre) to around 50 nanometres over several decades, and there is still some life in the existing processes.

The number of transistors on a single chip will shortly pass a billion and may reach 10 billion the next decade. Memory sticks will hold as much data as today's hard-drives. Eventually the chip-makers will hit a limit imposed by the way energy and matter behave at the smallest scales, but that has been predicted for some time and we are not likely to reach it in the next decade.

There will be a steady, perhaps sudden, disappearance of the venerable keyboard as our method for accessing our 'information devices'. Voice-recognition technology has advanced enormously in the last decade, as anyone booking a taxi nowadays will know. The software is smarter, taking account of not only speech sounds but also the structure of the language and even the user's verbal idiosyncrasies. Added to that, much more processing power and data storage is available even in small devices. Within a decade we are likely to be talking in plain language to the many microprocessors in our lives and they will talk back.

Driven by advances in optic fibres, lasers, LEDs and other optical devices, IT will be increasingly transformed by 'photonics', the use of light in place of electric currents and charges to process, store and transmit information (we are doing the latter of these now). This will continue the twin trends of lower costs and greater capacity, and may even postpone the end of the Moore's Law era. There will be progress in 'quantum computing', which promises to change IT beyond recognition, but we are unlikely to see a practical affordable device in the next decade.

Supercomputers will continue to get faster, with many benefits, especially in the challenging areas of weather and climate forecasting. Current technology has pushed the weather forecast periods out to ten days. The inherent unpredictability of the way the atmosphere/ocean system works to generate weather (for example, intense lows live for only seven to 10 days) will limit any longer range forecasts, but existing forecasts will be more reliable. Sorting out the human impacts on global climate depends heavily on supercomputing. Ten years should greatly narrow the existing uncertainties.

Measuring instruments in orbit around the Earth will continue to monitor continuously and in detail the state of our planet's health in terms of climate, pollution levels and ecosystems (more IT, really). Having such information does not mean that the necessary preventive and remedial steps will be taken, but an effective response is impossible without it. We will continue to explore the solar system using robotic spacecraft of increasing sophistication, as we have for more than 40 years, and will likely find more surprises.

Biotechnology

Biotechnology will continue to grow in importance in medicine, agriculture and environmental management. (Biotechnology is arguably a form of IT, since it operates by changing the information coded in an organism's genes.) Nanotechnology, which deals with objects the size of single atoms or molecules, is starting to make its mark. Some foresee a coming together (a convergence) of IT, biotech, nanotech and cognitive sciences, but the fruits of that are someway off yet.

The momentum for genetic modification (GM) of plants, and perhaps of some animals, for food and fibre providing a range of desirable qualities not available in nature, will continue, driven by the major investments that corporations in the field have already made. The debate over GM, such as the role of regulation, will also continue, hopefully enhanced by better access to information provided by the Internet.

Medical Treatment

Medical technology for prevention, diagnosis and therapy will continue to advance, driven by established sources of innovation like biotechnology, surgical techniques, monoclonal antibodies, medical electronics and nuclear medicine (the use of radioactive materials). We are already seeing the fruits of that in increasing life spans and healthier old age, though the aging of the population is setting a quick pace on the demand side. Implantable hearts, pacemakers and hearing devices may well be joined by implantable artificial kidneys, pancreases (to deal with diabetes) and vision aids.

Robots will become more common in the operating theatre, and expert systems (specially programmed computers that can undertake routine diagnosis) will find a place in GPs' offices. Those advances, aided by better communications, will advance treatment in remote areas and, we hope, developing countries. High on the list of targets will be effective therapies for global diseases like malaria and AIDS. There is already promising progress.

Transport

Both global and local transport will continue the current mix of modes. We are not likely to see any advance comparable to the mass production of automobiles, the introduction of high-speed trains or the inauguration of jet airliners. No doubt we will see incremental changes reducing costs and environmental impacts and increasing safety and comfort (say, through more hybrid vehicles), but the long lead times for development and the massive investments needed indicate that what will affect us over the next decade is already in the pipeline.

Energy

Energy will continue to be an urgent issue. How do we provide sufficient energy to run or lives, our society and our economy, while at the same time limiting the damage to our environment, particularly through the production of 'greenhouse gases'? We need both demand and supply-side solutions, and many technical options are already in view. Demand can be moderated by using energy less wastefully, for example, more hybrid vehicles and more fluorescent lamps or LEDs for lighting.

Photovoltaics and fuel cells can help on the supply side, as would the development of a way to produce hydrogen from water and sunlight, perhaps using bacteria. However, the massive capital sunk in the existing infrastructure based on coal, oil and natural gas, together with the capital that would be required to implement the alternatives, means we will see no major shift in the pattern of energy supply for a decade or more. There will be substantial testing of the concept of 'carbon capture' and sequestration as a way to reduce the global-warming impact of fossil fuels. New coal-fired power stations are likely to use more advanced technologies (already known) to extract more energy from the coal.

These predictions are conservative. Most are not likely to be fulfilled in the next five years, let alone ten. No doubt even more exciting things lie further down the road, but the path to them and the timing is less certain.

I have not postulated any major new understanding of the way the world works that can be turned into technology, nothing comparable to the discovery of latent heat that so improved the efficiency of steam engines; of electromagnetism and electromagnetic induction that underwrote the electric age; or of nuclear fission that found application in both peace and war. In the domain of everyday objects, systems and forces, centuries of research have probably left few surprises, but we cannot be sure. If, for example, we could show that superconductivity (total loss of electrical resistance) or even nuclear fusion can be made to occur at room temperature (and there is some possibility of both), practical applications could follow quickly.

What will drive technology forward (and make extra entries for future editions of this book) will be the blending of human need with human ingenuity and persistence. As this book shows, that combination liberates invention and innovation and changes lives.

Information Sources

Virtually all of the information used in writing this book has been obtained via the Internet/World Wide Web. This immense and expanding source of documents is easily and quickly accessible and commonly much more up-to-date than printed sources. However, its use requires some caution. Typing the name of an inventor or an invention into a search engine will generate almost instantaneously hundreds or even thousands of responses. Given the ease of 'cutting and pasting', the majority of these will be derivative, reproducing without significant editing and usually without any acknowledgement material available elsewhere.

It is therefore crucial to assess as much as possible the originality and reliability of the information presented. I apply a number of criteria here. One is the independence of the source. Many of the responses will be from websites with a commercial association with the invention. While these are often sources of detail not available elsewhere, their vested interest may lead to some imbalance in the way information is presented. More broadly, I look for significant differences in presentation or detail between rival sources to ascertain whether they are independent, and therefore can be used to crosscheck information or whether one is derived wholly or in part from the other.

Much of the information supplied is in the form of anecdote; stories that have been passed down and are not based on any original document. These give a lot of enjoyable flavour to the accounts of inventors and inventions in this book but their reliability may be questioned. We commonly read slightly varying versions of these stories, with no real way of knowing which if any is most reliable. My approach here has been to acknowledge the nature of this material, and preface it with phrases like 'one version of the story says that…'

One of the most widely used, and in my judgment reliable online resources, is **Wikipedia (http://en.wikipedia.org/wiki/Main_Page)**. This is a remarkable and astonishingly diverse and comprehensive collaborative online encyclopaedia, established with protocols designed to build confidence in the accuracy and objectivity of material published. Entries are commonly compiled what by individuals who appear to have a close association with and detailed knowledge of information presented and are replete with embedded links.

Wikipedia provides alternative ways to reach particular information, such as listings of inventors at **http://en.wikipedia.org/wiki/List_of_inventors**, and of inventions named after individuals at **http://en.wikipedia.org/wiki/List_of_ inventions_named_after_people**. The article 'History of Technology' (http://en.wikipedia.org/wiki/History_of_technology) is useful and comprehensive, as is the 'Timeline of Inventions' (http://en.wikipedia.org/wiki/Timeline_of_inventions).

Another significant source dealing with inventions and inventors forms part of the comprehensive **About** site (http://about.com). Listings of inventors and inventions arranged alphabetically can be found on http://intentors.about.com, with short informative pieces and hyperlinks to other sources. There are also useful and reasonably comprehensive timelines, again with embedded links to information sources.

The sources at **Ideasfinder** (http://www.ideafinder.com/history/index.html) again list by inventors, invention and times, and likewise at **Factmonster** (http://www.factmonster.com/ipka/A0004637.html). However, the need to assess the reliability of information found by the various routes, as discussed above, remains.

As to printed resources, I found good value in the *Timetable of Technology* (Hearst Books, New York, 1982), dealing with twentieth-century inventions, and *Eureka: An Illustrated History of Inventions from the Wheel to the Computer*, edited by Edward de Bono (Thames and Hudson, London, 1979), though both of these are obviously deficient in coverage of recent decades.

Fascinating perspectives on how important new technologies were viewed much closer to their time of invention have come from books like *The Wonderful Century* by A.R. Wallace (E.W. Cole, Sydney, 1906), *Science Remaking the World* by Otis Caldwell and Edward Slosson (Garden City Publishing, New York, 1923), *The Science Year Book for 1945*, edited by John Ratcliffe (Doubleday, Doran and Company, New York 1945) and *A Treasury of Science,* edited by Harlow Shapley, Samuel Rapport and Helen Wright (Angus and Robertson, London and Sydney, 1954).

My understanding of the web of historical contexts and parallel events has been enriched by books like *Phillips World History: People, Dates and Events* (George Phillip Limited, London, 1999) and a much loved and all-but-disintegrating *Encyclopedia of Dates and Events,* edited by L.C. Pascoe (Hodder and Stoughton, London, 1968). J.D. Bernal's iconic *Science in History* (Penguin Books, Melbourne, 1969) is a constant companion.

David Ellyard

Timelines

The following lists group the 250 stories in this book into their main areas of application, while keeping the chronological order. This will help you work your way through developments in the areas of technology that most interest you.

COMMUNICATIONS

1565 Something to Write With
Quill Pens, Pencils, Fountain Pens

1792 Information Outruns the Horse
Claude Chappe

1799 Better Paper, Faster Printing
Nicholas Robert, Friedrich Koenig, William Bullock

1827 Now the Blind Can Read
Louis Braille

1837 Messages Over the Wires
William Cooke, Charles Wheatstone

1840 Paying for the Post
Rowland Hill

1844 The Telegraph Expands
Samuel Morse

1858 The Telegraph Goes Global
William Thomson (Lord Kelvin)

1868 Machines that Write
Christopher Scholes

1876 The Birth of the Telephone
Alexander Graham Bell, Elisha Gray

1887 Inventing a Language
Ludwik Zamenhof

1896 The Telegraph without Wires
Heinrich Hertz, Edouard Branly, Guglielmo Marconi

1906 Music in the Air
Reginald Fessenden

1908 Pioneers of Television
Shelford Bidwell, Paul Nipkow, Alan Campbell-Swinton

1920 The Birth of Radio Broadcasting
David Sarnoff

1933 A Life for Radio
Edwin Armstrong

1938 New Ways to Write (and To Correct)
The Biro Brothers, Bette Graham

1963 Communicating by Satellite
Arthur C. Clarke

1965 Optic Fibres Transform Communications
Charles Kao, George Hockham and Others

1971 A Foretaste of the Internet
ARPA, Ray Tomlinson

1973 Talking while Walking
Martin Cooper

1991 The World Wide Web
Tim Berners-Lee

ENERGY AND LIGHTING

1650 Can We Use the Power in the Air?
Evangelista Torricelli, Otto von Guericke

FOOD AND FIBRES

GADGETS AND DOMESTIC TECHNOLOGY

INFORMATION TECHNOLOGY

THE TECHNOLOGY OF KNOWLEDGE

MATERIALS AND MANUFACTURING

1978 The Girl from the 'Test Tube'
Patrick Steptoe, Mu Chueh Chang
Robert Edwards, the Monash IVF Team

1985 The Ultimate Hearing Aid
Graeme Clark

1996 A Clone Called Dolly
Hans Spemann, Ian Wilmut

RECREATION AND ENTERTAINMENT

1564 The Technology of Music
Andrea Amati, Adolph Sax and Many Others

1709 Keyboards that Play Loud and Soft
Bartolomeo Cristofori

1787 A Stick and a Ball
Cricket, Baseball, Tennis and Other Games

1863 Kicking a Ball Around
The Many Forms of Football

1863 Get Your Skates On
James Plimpton, the Olsen Brothers

1884 All the Fun of the Fair
George Ferris, John Miller

1895 Machines that Make Music
The Pianola and the Jukebox

1895 The Moving Picture Show
Thomas Edison, the Lumière Brothers

1913 The Games People Play
Arthur Wynne, Clarence Darrow

1927 The Movies Learn To Talk
The Legacy of The Jazz Singer

1928 Reinventing Ancient Toys
Pedro Flores, Donald Duncan, Richard Kneer, Arthur Melin

1931 Fun in the Arcade
Raymond Maloney, Harry Mabs

1934 Electronics Makes Music
Laurens Hammond, Léon Theremin

1935 Movies in Colour
Herbert Kalmus

1936 Television Goes on Air
Vladimir Zworykin, Philo Farnsworth

1943 Breathing Underwater
Jacques Yves Cousteau, Emile Gagnan

1994 More Music in Less Space
Music Compression Becomes Mainstream

1996 More Movies in Less Space
The DVD Triumphs

THE BIG PICTURE

1513 A Most Ingenious Mind
Leonardo da Vinci

1626 'The Effecting of All Things Possible'
Francis Bacon

1752 A Man of Many Parts
Benjamin Franklin

1863 Visions of the Future
Jules Verne, H. G. Wells and Others

1884 The High Voltage Wizard
Nikola Tesla

1900 The Wonderful Century
Alfred Wallace Sums Up

Index of Inventors

Index of Inventions